I0468992

STRATEGIES FOR MARKETING YOUR FIRE DEPARTMENT TODAY AND BEYOND

Federal Emergency Management Agency

United States Fire Administration

CONTENTS

Strategies for Marketing your Fire Department Today and Beyond

Local children participate in the Department's Safe Center for Children activities.

Fairfax County Fire and Rescue Department Public Information Officer provides incident information to news media.

Firefighter instructs citizens on proper fire extinguisher use.

This publication was produced under contract EME-96-CO-0236 for the United States Fire Administration, Federal Emergency Management Agency. Any information, findings, conclusions, or recommendations expressed in this publication do not necessarily reflect the views of the Federal Emergency Management Agency or the United States Fire Administration.

Photos courtesy of Fairfax County Fire and Rescue Department

ACKNOWLEDGEMENTS

The United States Fire Administration (USFA) would like to thank the following for their contributions to this manual:

Ed Kirtley
Chief
Guymon, Oklahoma, Fire Department

Carol Gross
Fire and Life Safety Program Manager
Phoenix, Arizona, Fire Department

Mary Nachbar
Bureau Chief
State Fire Marshal Division
Minnesota Department of Public Safety
St. Paul, Minnesota

Michele Frisby
Deputy Director/Chief Information Officer
Communications and Information
International City/County Management Association
Washington, DC

Sheretha Jimeno
Kindergarten Teacher and Learn Not to Burn ™ Instructor
Poinciana Park Elementary School
Miami, Florida

Joseph Lowe
Training Captain
Orange County, California, Fire Authority

Mary Beth Michos
Chief
Prince William County, Virginia Fire and Rescue

Kenneth H. Stewart
CEO
Empowered Learning Incorporated
Kimberton, Pennsylvania

Raymond E. Hawkins
Director
Client Education and Training Services
Volunteer Firemens Insurance Services, Inc.
York, PA

Ken Knipper
Disaster and Emergency Services Director
Campbell County, Kentucky and
National Volunteer Fire Council Kentucky Director

Susan McHenry
EMS Specialist
National Highway Traffic Safety Administration
Washington, DC

Andrea A. Walter
Program Manager
International Association of Fire Chiefs
Fairfax, VA

CHAPTER 1   The Vision

"It ain't like it used to be!" The old cliché certainly is appropriate for today's fire service compared to yesterday's. But, what about tomorrow—what will the fire service of the 21st century be like? The answer to that question is: It will be a fire service filled with opportunities and challenges.

The future of the fire service is being, and will continue to be, driven by changes in society. These changes include new expectations of the citizens served by the fire service—such as greater accountability for the use of public resources and improving the overall efficiency of programs and services. Members of the fire service have new expectations. Also, the fire service will continue to become more diverse, in both members and services.

> *"In the beginning of the next century the fire service's mission will continue to evolve into a more broad-based risk control and management organization. The typical fire organization will deliver a wide range of emergency and non-emergency services to deal with topics such as seat belt safety, out-of-hospital health care, and other risks/ hazards faced by our society."*
> — Bill Peterson, Chief, Plano, Texas, Fire Department.[1]

To keep pace with society, the fire service must review its current mission and determine if that mission will indeed meet the demands of the next century. The old mission of simply "saving lives and protecting property" may no longer have the depth or scope necessary to meet the expectations of the public. If, indeed, the mission must change, that change should come from within the fire service. This will require innovation, courage, and the commitment of fire service leaders at all levels, both career and volunteer. In fact, change may be the single greatest challenge facing the fire service in the next century.

The purpose of this manual is to assist fire service leaders in examining the future, the role of the fire service in that future, and ways to "get there from here." It is designed to provide a fire chief, a public information officer, and other leaders in the fire service with guidance and tips on marketing a department and its services to the local customers: the citizens and organizations served by the department.

New Mission

> *"Innovation requires us to systematically identify changes that have already occurred—in business, in demographics, in values, in technology or science—and then to look at them as opportunities. It also requires us to abandon rather than defend yesterday—something that is most difficult for existing companies to do."*
> — Peter Drucker, Chair of the Drucker Foundation and prominent management consultant.[2]

Success in the 21st century will require new ways of doing things. The management methods and programs of today's fire service may work in 1997, but may not hit the mark in 2000 and beyond. A new philosophy will be needed for the dynamic new environment in which the fire service will operate. That philosophy involves an expanded mission for the fire service.

The traditional mission of the fire service was once very clear: saving lives and protecting property from fire. Today, however, that mission has taken on a much broader meaning. Fire departments today do much more than simply suppress fires. In the past 25 years, fire departments have tackled the problems of hazardous materials, developed effective emergency medical services to provide treatment to the sick and injured, implemented comprehensive community education programs designed to reduce the frequency of preventable fires and injuries, and have even begun to provide neighborhood health care services, including vaccinations.

In the future, the fire service's mission will still include saving lives and reducing property loss. But it will include more, much more. According to Chief Alan Brunacini of the City of Phoenix, Arizona Fire Department, the mission of the fire service in the future will be to "prevent harm." Preventing harm will cover many areas, including specialized rescue, health and wellness of citizens, and injury prevention. Protecting property will include protecting community resources—people, property, natural resources, the environment, and the community infrastructure—from harm and loss. Also, protecting property will include mitigation of natural and technological disasters.

The most important part of the new mission will be how it is accomplished. For generations the primary tool the fire service used to save lives and protect property has been emergency response. While response will always be a key function of the fire service, in the next century there will be a focus on prevention of emergencies. This emphasis on prevention is a fast-growing paradigm in society. Simply put, preventing fires, injuries, and disease is the most effective means of "preventing harm." In the future, public education and prevention will be of equal importance to fire suppression in the role of the fire service in the community.

Prevention programs and services are not new to the fire service, but in the years to come these tools will grow dramatically in both scope and priority. The prevention services of the future will include more comprehensive community education programs in injury prevention, fire prevention, and personal health care; using technology to improve fire protection systems; adopting of more effective life safety codes; and using operational personnel to conduct neighborhood education and presentations.

It is important to look at the relationship of public information, public education, and public relations functions and their role in fire departments, both today and tomorrow, especially since they are keys to effective prevention programs. There is often confusion about each function, since they are similar, but have some differences. Together, the activities that are used to accomplish these three functions are known as PIER programs (public information, education, and relations).

Public information is the process of informing the public about the operations of and actions taken by the fire department during emergencies. The public has a right to know about these operations, and public information ensures public awareness about the emergency services provided by the department.

The public information function generally is accomplished through the media. Public information can also be provided through public speeches and other presentations.

Public education is the process of changing people's attitudes and behavior related to safety, as most fires and injuries can be prevented with changed behavior. Think about how many fires and injuries are due to a person's behavior: food left on the stove, ashes placed in a pasteboard box, riding an ATV without a helmet, matches left on the table within reach of a young child. These incidents are due to careless or inappropriate human behavior. Public education seeks to change a person's attitude about his or her personal safety and wellness, resulting in a change in behavior.

Public education uses many methods to achieve its goal, including presentations at schools, programs at community events, news stories, and public service announcements.

Public relations is the process of developing a positive public perception about the fire department, its members, its programs, and its services. Public relations is an active function and doesn't "just happen." Public relations programs use many methods to reach the public, including news shows, public service announcements, newspaper articles, fact sheets, presentations to the public, programs at community events, and the like. The focus of a public relations program is often the value of the programs and services provided, including cost effectiveness and improvement of community safety and wellness.

What do the three functions have in common? First, all are aimed at members of the public. Second, each function employs multiple methods, and often the same methods, to reach the public. Third, when one function is done well, it supports the other functions.

For example, when a department provides a quality public education presentation in a school, it also provides effective public relations. Parents learn that the department is working to teach fire safety to their children. Teachers learn that the department is supporting education efforts in the community. The result is that public education is also developing a positive public perception about the department.

PIER programs are essential for winning the support of a community's decisionmakers. Most decision makers do not understand fully the challenges facing fire departments, nor do they appreciate the scope of services generally provided to the community. This may be due to a lack of information about the department. Effective public information programs help decisionmakers understand the emergency services provided to the community. Public information may also enlighten decisionmakers about the fire or injury problems in the community.

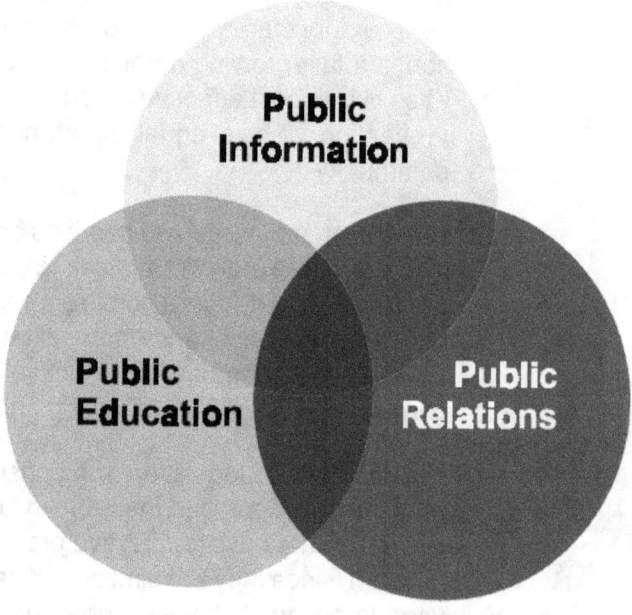

Figure 1, PIER Functions

Public relations programs build a positive perception within the community about the department. This positive feeling is usually communicated to the decision makers, sometimes through formal recognition of the department, and sometimes by word-of-mouth. In either case, public relations programs help decisionmakers understand that the customers are pleased with the fire department.

Public education programs help decisionmakers understand that there are solutions to the problems other than responding after the fact. Public education programs also show decisionmakers that the department is attempting to prevent problems proactively , a strategy that is cost effective and popular with taxpayers.

A PIER program is a powerful tool for the fire department. It helps to build public support for the department and helps prevent fires and injuries. To be successful in the future, fire departments will be using PIER programs more and more to reach the public.

Benefits of a PIER Program

Benefits to Fire Department

- Increases efficient use of existing resources by educating the public to prevent or properly respond to emergencies

- Increases political exposure and support

- Increases public support for additional resources

- Improves retention of current members

- Improves morale through positive public and internal recognition

Benefits to the Community

- Reduces the number of deaths, disabilities, and property loss by educating the public on prevention, self-care, first response, and bystander care

- Reduces costs (for the individual citizen and the community)

- Reduces emotional trauma

- Improves service

- Increases community pride, which aids in recruiting new businesses and residents

Benefits to You

- Increases recognition and rewards for community service

- Increases job security and ability to move up in ranks (more and more fire departments are creating full-time PIER positions)

- Increases personal satisfaction

- Improves the working environment

"Our mission in the next century will certainly place as much, or more, emphasis on preventing loss, including injuries, as on simply responding to the emergencies. This will include code enforcement, fire and life safety education, and fire protection systems."

— Ed Kirtley, Training Director, Colorado Springs, Colorado, Fire Department and Chair of the NFPA 1035 committee for Fire and Life Safety Educator.[3]

A View of Tomorrow

The expanded mission of the fire service will be driven by changes in society, which will mold and shape the programs and the people in the fire service.

Proliferation of Information

Knowledge is expanding rapidly, making it impossible to stay abreast of all the new information that is available.

The technology that is available to communicate this information is also growing at an unbelievable rate. We now have the option of many different ways of communicating with someone else: face-to-face; in writing; via telephone or cell phone from practically anywhere; by facsimile, modem, or e-mail; and via the Internet. Information transfer that used to take days can now be completed in seconds.

Technology will have a dramatic impact on a fire department's ability to gather, analyze, and disseminate information, both internally and externally. It will also have a significant impact on the way in which personnel are trained.

Slow Decline of Metropolitan Core

Metropolitan areas continue to grow, but the benefits of this growth are not shared by all. The middle class continues to move to the suburbs, which continue their inexorable spread. The central core of the city continues its slow decline. This decline not only affects the condition of the buildings and the infrastructure, it also affects the quality of life for those left behind. Health problems, crime, and increased fire risk are all part of this process of decline. Unfortunately, as the tax base for the city erodes, the tax revenue needed to combat these problems also declines. Government agencies are sometimes challenged to meet even the most basic needs of the citizens.

Changing Demographics

This country is home to a great diversity of people. In the future, minorities will become the majority, in large part due to immigration from Central and South America and from Asia.

As baby boomers age, the proportion of older people within the population will increase. This segment of the population will require new programs to meet its unique fire and life safety needs.

> *"The larger number of older people will require specially targeted education. Many may live in new types of arrangements designed for the elderly that will have special fire and life safety requirements. Outreach appropriate to the lifestyle and concerns of this group will be needed."*
> — *Leadership in Public Fire Safety Education: The Year 2000 and Beyond.*[4]

These changes will challenge the fire service to find new and innovative ways to reach each group with programs that will meet its needs. It is also going to require fire departments to reach out aggressively into each neighborhood and community. One program for all people will not work in the future.

Focus on Prevention

Organizations that provide human services to the community will place a much greater emphasis on prevention. This will include educational programs designed to change behavior, codes and standards to mitigate risk, and analysis of incidents to determine more effective methods for preventing fires and injuries.

The private sector—principally health care insurance providers—will continue to focus on wellness. These efforts will be designed to reduce the cost of health care.

Government will also emphasize prevention with the implementation of managed health care programs. Such programs may have a dramatic impact on prehospital emergency medical care provided by fire departments.

> *"Increasingly, emphasis will be placed on local prevention of social and medical problems; 'cures' will be just too expensive."*
> — *Dr. Harold Hodgkinson.*[4]

Greater Competition for Scarce Resources

The budgets of most local governments have been strained in recent years. This will certainly continue in the future. Budget pressures will be influenced by many factors, including the condition of the local economy, the local political environment, and the credibility of the local elected officials. Local fire departments will feel the effects of this strain.

As financial resources grow tighter, the competition among government agencies for those resources will increase. Fire departments will have to compete with other departments for those funds. They will also have to compete with other city governments for scarce State and Federal funds. Even today, this competition is occurring between public fire departments and contract fire protection companies. These companies are providing contract fire protection services in numerous communities, and this trend may continue as more and more cities seek ways to reduce the cost of providing municipal services, including fire protection.

This will require fire departments to market their services effectively to decision makers as well as to the citizens of the community. They will have to demonstrate the value of their services and programs, and the benefit to the community from providing the needed budget.

Community Emergency Response Teams (CERT)

This fiscal pressure may place more onus for safety and protection on citizens themselves, resulting in a greater focus on "self-help" among citizens. In California, the emergence of Community Emergency Response Teams (CERTs) is a good example of citizens working together to help each other during emergencies. These programs are expanding to other areas of risk.

Community Emergency Response Teams (CERT)

Introduction

A major disaster which overwhelms a community's or region's resources will preclude first responders meeting the demand for fire and medical services. Factors such as the number of victims, communication failures, and road blockages will prevent people from accessing emergency services they have come to expect at a moment's notice through 911. People will have to rely on each other for help in order to meet their immediate life-saving and life-sustaining needs.

Under these kinds of conditions, family members, fellow employees, and neighbors will spontaneously try to help each other. This was the case following the Mexico City earthquake where untrained, spontaneous volunteers saved 800 people. However, 100 people lost their lives while attempting to save others. This is a high price to pay and is preventable through training. That is what the CERT program is about.

Background

The Community Emergency Response Team concept was developed and implemented by the Los Angeles City Fire Department (LAFD) in 1985. The Whittier Narrows earthquake in 1987 underscored the area-wide threat of a major disaster in California. Further, it confirmed the need for training civilians to meet their immediate needs. As a result, the LAFD created the Disaster Preparedness Division with the purpose of training citizens and private and government employees. As of 1993, more than 10,000 people, and over 267 teams have been trained.

The training programs that LAFD initiated make good sense and further the process of citizens understanding their responsibility in preparing for disaster. It also increases their ability to help themselves, their family, and their neighbors safely . The Federal Emergency Management Agency (FEMA) recognizes the importance of preparing citizens. The Emergency Management Institute (EMI) and the National Fire Academy (NFA) adopted and expanded the CERT materials, believing them applicable to all hazards.

The CERT course will benefit any citizen who takes it. This individual will be better prepared to respond to and cope with the aftermath of a disaster. Additionally, if a community wants to supplement its response capability after a disaster, civilians can be recruited and trained as neighborhood, business, and government teams that, in essence, will be auxiliary responders. These groups can provide immediate assistance to victims in their area, organize spontaneous volunteers who have not had the training, and collect disaster intelligence that will assist professional responders with prioritization and allocation of resources following a disaster. Since 1993, when this training was made available nationally by FEMA, communities in the States of California, Washington, Oregon, Florida, Missouri, Utah, Oklahoma, Tennessee, and Kentucky have conducted the training.

Starting

FEMA recommends a number of steps to start a CERT:

Identify the program goals that CERT will meet and the resources available to conduct the program in your area.

- Gain approval from appointed and elected officials to use CERT as a means to prepare citizens to care for themselves during a disaster when services may not be adequate. This is an excellent opportunity for the government to be proactive in working with its constituency.
- Identify and recruit potential participants. Naturals for CERT are community groups, business and industry workers, and local government workers.
- Train CERT instructor cadre.
- Conduct CERT sessions.
- Conduct refresher training and exercises with CERTs.

Delivery

The CERT course is delivered in the community by a team of first responders who have the requisite knowledge and skills to instruct the sessions. It is suggested that the instructors complete a CERT Train-the-Trainer (TTT) course conducted by their State Training Office for Emergency Management or the Emergency Management Institute in order to learn the training techniques that are used successfully by the LAFD.

The CERT training for community groups is usually delivered in 2-1/2 hour sessions, one evening a week over a 7-week period. The training consists of the following:

- Session I, DISASTER PREPAREDNESS: Addresses hazards to which people are vulnerable in their community. Materials cover actions that participants and their families take before, during, and after a disaster. As the session progresses, the instructor begins to explore an expanded response role for civilians, in that they should begin to consider themselves disaster workers. Since they will want to help their family members and neighbors, this training can help them operate in a safe and appropriate manner. The CERT concept and organization are discussed, as well as applicable laws governing volunteers in that jurisdiction
- Session II, DISASTER FIRE SUPPRESSION: Briefly covers fire chemistry, hazardous materials, fire hazards, and fire suppression strategies. However, the thrust of this session is the safe use of fire extinguishers, sizing up the situation, controlling utilities, and extinguishing a small fire.

Conclusion:

CERT is about readiness, people helping people, rescuer safety, and doing the greatest good for the greatest number. CERT is a positive and realistic approach to emergency and disaster situations where citizens will be initially on their own and their actions can make a difference. Through training, citizens can manage utilities and put out small fires; treat the three killers by opening airways, controlling bleeding, and treating for shock; provide basic medical aid; search for and rescue victims safely; and organize themselves and spontaneous volunteers to be effective.

Additionally, CERT training provides a valuable opportunity for the emergency response community to educate the public about itself and the services it provides. CERT is an excellent marketing tool. Finally, the program creates a relationship between the response community and the people it serves, resulting in grassroots support for response organizations.

For more information, please visit the Emergency Management Institute Web site at
http://www.fema.gov/emi/cert
or write the Emergency Management Institute at 16825 South Seton Avenue, Emmitsburg, Maryland 21727

Another result of budget pressures will be the formation of more collaborative agreements with other public and nonprofit agencies and public-private partnerships with private sector businesses. These agreements will help create new programs and continue existing ones. A good example of this type of partnership is the Learn Not to Burn ™ Champion grant program sponsored by the National Fire Protection Association (NFPA). Departments chosen for this grant receive financial assistance and training support for conducting the Learn Not to Burn ™ program in local schools.

Partnerships may include getting funding from businesses and industry for public education programs, sharing training facilities with local colleges, and sharing communication facilities with other public safety agencies. Fire departments may partner with health departments and hospitals to develop and implement programs designed to reduce fires, injuries, and even illness. By combining these scarce resources, the partners will be able to provide effective community-based programs.

> *"Our imaginations in creating the blueprints will outdistance our resources to actually build the solutions. . . Creating and maintaining strategic alliances that work in an intelligent, orchestrated fashion will help us apply our limited resources where they'll make the most difference."*
> — *Meri-K Appy, Vice President, Public Education, National Fire Protection Association.*

Citizens Will Demand More Effective Government

Once upon a time, citizens did not demand accountability from the fire department. That is all changing, as citizens demand more accountability of all segments of their local governments. Taxpayers demand to know how their tax dollars are being spent, and they want to know if there are better ways of providing the same government services. Fire departments must now demonstrate the need for services, and must clearly show that they are doing everything possible to provide the services and programs effectively and efficiently.

When fire departments can't live up to this scrutiny, budgets may be decreased. In some cases, budget reductions may result in layoffs and station closures. In other cases, local elected officials may choose to contract with private firms to provide services and may simply eliminate or downsize the fire department. In rural areas, local and county governments may require volunteer departments to raise a greater percent of their budget through community fundraising activities.

This trend will force the fire service to pay more attention to the needs of the citizens. In other words, fire departments will have to commit to effective customer service. By identifying the needs of the customers, and then developing cost-effective programs and services to meet those needs, fire departments will demonstrate to taxpayers that they are good stewards of the local tax dollar.

> *"Fire departments will need to be on the lookout for more opportunities to serve the public. This process will be based on the competitive nature of obtaining funding. Communities are holding elected and appointed officials accountable for their actions and productivity. Departments that cannot and will not meet additional needs for the public that fit within the mission of the fire service will be replaced by commercial operations."*
> — *Dennis Rubin, Chief, Dothan, Alabama, Fire Department.[6]*

Achieving the Mission

The expanded mission of the fire service will certainly be filled with challenges. New programs will be needed, but there will be less revenue available to provide the programs. A fire department will be required to prove the value of its programs and services to the taxpayer. This proof will have to be provided in understandable, tangible terms. In a real sense, the mission of the fire service in the next century goes beyond preventing harm. The mission of the fire service may be to simply survive this new environment.

Adapting to the new and changing environment will require the fire service to re-create itself, a process begun already by many departments. The fire service will have to focus on fiscal effectiveness, expansion of programs, strategic partnerships within the community, and most importantly, determining and meeting the needs of the customer.

To do this, leaders in the fire service must have a plan. They must consider where they are today in terms of programs and services. They must identify current and future customers. They must develop a roadmap to the future.

Fire department leaders will have to promote their vision of the fire service of the future to anyone and everyone who will listen: elected officials, fire department employees, and customers. The product of these efforts will be a healthy, viable department ready to meet the challenges of tomorrow, regardless of what they may be. The information in this manual is intended to help you, the fire service leader, achieve this mission.

> *"The continued existence of the fire service as we know it will hinge on the organization's ability to meet the wants, needs, and expectations of our society."*
> — Bill Peterson, Chief, Plano, Texas, Fire Department.[1]

The benefits from public information, public education, and public relations may seem unattainable, especially for departments that are struggling already to provide adequate services and programs. Yet, the availability of local resources many times depends on the public's perceptions of the fire department. When the public believes the department is providing cost-effective, valuable services, resources are easier to obtain.

The following case study illustrates the positive impact of a public information, education, and relations program on a small fire department.

PIER Program's Effect in Salida, Colorado

Salida, Colorado is a small mountain community of approximately 5,000 people. Its primary industries are tourism and ranching. It has a combination fire department. Until 1995, the full-time staffing consisted of two firefighters, two captains, and a chief. In 1995, a new chief was hired.

On his arrival, the new fire chief, Peter DeChant, found a fire department with a poor community image. DeChant talked to many citizens about their perception of the fire department. The consensus was that the fire department did nothing, that its members stood in front of the station and tried to look important, that they did not know how to fight fire, that they looked like slobs, and that they needed to be shaped up. The same perceptions were expressed by city council members and the city administrator. DeChant found the morale of the department members equally negative. In short, the department needed help.

At that time, the department did not have any organized public education, information, or relations programs. In fact, the department minimized contact with the public.

DeChant's first step to change the department was meeting with department members to formulate expectations and a strategy for change. They established goals, including improvement of customer service and public relations, more involvement with the community, and creation of a public education and prevention program.

The department purchased new uniforms, including badges, name bars, and collar brass, to improve the members' appearance. Also, the station was cleaned so that it would be a showplace for the community.

DeChant immediately implemented a positive public relations strategy. He began speaking at every public event possible, including meetings of civic clubs, neighborhood groups, and radio talk shows. He invited all citizens to come by the station and meet the firefighters and learn about the department.

A new public information strategy was also implemented. DeChant encouraged the local media to learn about the fire department. Any information on the department or on incidents handled by the department was provided quickly to the media. He didn't wait for the media to call for the information, he ensured that the information was sent to the media immediately. More and more information about the department began reaching the public.

Public education also became a priority. The department began a free smoke detector/battery replacement program. It provided fire extinguishers and fire safety presentations whenever requested. First aid and CPR classes were offered to the public. The fire department joined forces with the police department to provide fire safety presentations through the community watch program. A fire safety program was implemented with the local hotel/motel lodging association. The program was so successful that the association requested that the fire department inspect all lodging facilities in the town.

The changes did not only involve PIER programs. The department began responding to emergency medical incidents and specialized rescue incidents. Training was increased. Professional requirements, for both the career staff and volunteers, were implemented.

By the end of 1996 the results of the new programs were clear. While calls had increased 108% due to expanded services, the actual fire loss in dollars decreased by nearly 70%. The city council, excited about the new changes, approved five new career firefighter positions. New equipment was purchased. In 1997, a fire marshal position was added, resulting in a 120% increase in career staff in two years.

Why was the Salida Fire Department able to make such gains in staff, reduce fire loss, and improve public perception? The answer according to Chief DeChant is simple. All of the fire department's members made a commitment to the community. Programs involving all members of the department moved the department and its programs from a negative perception to a positive perception in the eyes of the public. This was accomplished through personal contact, the media, and public education presentations. The proactive strategy for gaining the public's trust and respect was tremendously successful. It was done with few resources or additional costs. It was accomplished through teamwork, hard work, and ingenuity, something available to every fire department.

Thanks to Chief DeChant and the members of the Salida Fire Department for sharing their story.

"As we move into the 21st century, there is scant evidence that the nature and pace of change is likely to slow. We will evolve in the 21st century and beyond in ways not now fore-seeable."
— Alan Greenspan, Chair, Board of Governors of Federal Reserve System.[7]

Change. It is certainly not something new to the fire service. Think about it. In less than 100 years the fire service has evolved from using horses and hose carts to using complex machines capable of pumping thousands of gallons of water a minute. Fire fighters no longer are forced to breathe deadly smoke and gases when they fight fires. Also, instead of only fighting fires, the fire service now provides a whole range of services to its customers.

The expansion of emergency services provided by the fire service has been dramatic in the past 20 years. In the late 1970s and early 1980s, there was prolific growth in the emergency medical services provided by fire departments. In the 1980s, local fire departments placed a new emphasis on hazardous materials response. The 1990s have been a time for growth in specialized rescue, including urban search and rescue operations. There has also been increased attention on responding to terrorist incidents because of the bombings in Oklahoma City (Murrah Federal Building), and New York (World Trade Center).

Throughout these past two decades, fire service leaders were keenly aware of the need to keep pace with changes in the world. They recognized that the fire service had to change to meet customer demands, and had to mature to compete for public resources. In short, fire service leaders realized that change was an essential part of the fire service's formula for success.

Today, the need for change is even more pressing because of both the changes occurring in society and the needs of local customers. Customers want a fire service that is fiscally efficient, but at the same time they want the same or more services. In some cases, new services are driven by the desires of the public, and in other cases, the services are needed to ensure adequate fire protection and life safety. There is also the competition from other government agencies, and the challenge of competition with private companies.

These pressures are not unique to the U.S. fire service. The fire service in other parts of the world is also faced with these same challenges. Brian Robinson, the Chief Fire Officer of the London Fire Brigade, sees the facilitation of change in preparation for the future as his greatest concern. Explains Chief Robinson: "As we move toward 2000, it is reasonable to expect organizations to ask questions about what changes they face if they are to provide effective services into the next century. This is perhaps even more important for a public emergency service such as the London Fire Brigade, and it has been my most pressing concern since I became chief fire officer in October 1991."

If change is inevitable, what can the fire service do to ensure that the change takes a positive direction? There are many steps that leaders in the fire service can take to ensure the inevitable change produces a healthy, viable fire service. However, it is essential to understand that change cannot be truly controlled. Change, both now and in the future, can only be guided, and the effects of the change only managed.

Fire service leaders must understand that change that is created or initiated from within the organization generally can be managed most effectively. It is also the least painful to the members of the organization. To this end, fire service leaders must be able to initiate and guide the change process within their fire service organizations. This skill will be critical to survival as a chief executive officer.

> *"The organization must be trained and prepared for a never-ending change process to be able to survive and grow. Change is not new. How well an organization deals with and manages change is the real issue."*
>
> — Dennis Rubin, Chief, Dothan, Alabama, Fire Department.[6]

Paradigms

To understand the process of initiating and guiding change within the organization, it is important to start with the concept of paradigms. One definition of a paradigm is a "model". In this case, it is a model that guides a person's behavior. In other words, each person has specific models or patterns that determine how he or she will react to any given situation. These paradigms are learned through experience, and are greatly influenced by the culture in which the person lives and works.

For example, a new firefighter in a department may be taught that the best way to attack a fire is using strictly a 1-1/2-inch handline. This is what the department expects. Over time, that firefighter will develop a belief this is a "normal" way to fight a fire. The model paradigm of using only 1-1/2-inch attack line has been learned.

The same process for developing a positive paradigm about PIER can also occur. A rookie joins a fire department to fight fire and save lives. This particular fire department strongly believes in the value and benefits of PIER programs. Throughout rookie training, and then as a member of the department, the new firefighter sees the veteran members of the department living this belief in PIER through their attitude towards the community, the programs provided, and their actions. Again, over time, the firefighter develops a belief, a paradigm, that public information, education, and relations programs are part of what the fire department does, are something that is expected of every member of the department, and are vital to the department's mission.

Organizations can also have paradigms. Each fire department is unique. Each has its own set of operational guidelines. Each has its own values and beliefs about such things as fireground strategy and professional development. The fire service professional also has paradigms that guide behavior. For example, respect shown chief officers by subordinates is a pattern that is generally seen throughout the fire service.

New paradigms—new models or patterns—are emerging and more will emerge in the future. These paradigms are forming as part of the response to the societal changes discussed in Chapter 1. A few of these new paradigms are described below.

Focus on Meeting Customer Needs

> *"Customer-centered means that customer needs, perceptions, and feelings begin to design and drive how the service delivery systems looks and behaves.... We have always done the very best we could for*

our customers, but we haven't spent much time asking them what they really want... simply, we decided what we thought they really needed, delivered that service, and went home."
— Alan Brunacini, Chief, Phoenix, Arizona, Fire Department.[8]

In the future, every fire service organization will have to work to identify customer needs, and then do everything possible to meet those needs. This will not be simply a fad and then fade; it will become an expected level of performance. Anything less will not be tolerated by the customer. In short, serving the community through effective, quality customer service will be an essential part of the fire service.

This model also applies to the internal customers of the fire department—the men and women who are the organization. They are customers also, and have needs and expectations that must be met. It is easy at times to think of these men and women as simply "employees" or the "rank and file." However, they are the driving force of the organization, and without them it would not be possible to deliver any programs or services to the citizens. Without satisfied internal customers, the needs of the external customer—the citizens in the community—will not be satisfied. Thus, fire department leaders have two groups of customers. Fire service organizations that quickly learn this paradigm will reap many benefits.

First, by learning and meeting the needs of the customers, organizations will help ensure that precious and limited resources are being applied to the areas of the greatest need. Keep in mind that the task of learning the needs of the customers requires the organization to constantly seek out each customer group—such as the elderly, residents of older neighborhoods, recent immigrants—and ask them for input. This is not always easy to accomplish. Also, the organization must be willing to act on the input. This will require the organization to be adaptable. By putting this into practice a fire department will use resources more cost effectively. And, most importantly, the customers' needs will be met. In addition, responding directly to the needs of the citizens is probably the best way to gain the support of the local appointed and elected leadership—the decisionmakers and controllers of the resources. Remember that the local appointed and elected leadership make the ultimate decisions on behalf of your customer.

By gaining the trust of the citizens and the decisionmakers, the chief will be able to bring forward new programs and needs that are considered objectively. The public will trust the judgment of the chief, and will be willing to help facilitate the needed change.

Second, by serving the internal customers, the fire department administration will improve the effectiveness of each individual and of the overall organization. Chief executives should be aggressive in seeking out the input and feedback of the men and women who work with the external customers. More than anyone else, the employees are aware of what is needed within the organization to develop success. Using their ideas and inputs, and being dedicated to maximizing working conditions, will create a healthy environment and will result in excellent service to the external customers.

Third, by being a committed advocate for the needs of both the external and internal customers, the fire chief will develop a critical element: the trust of those served. This trust, as much as anything else, will be essential for creating changes in a fire service organization, and in the community. For internal changes, the chief must have the trust of the personnel. The personnel must believe that the chief is acting in their best interest, and that the changes will improve the department. Without this trust, the personnel may fight the change through either apathy or organized resistance to the new program or process.

Finally, dedicating the resources of a fire service organization to learning and then meeting the needs of the external customers results in a high level of community equity. Community equity is similar to money that a person puts into a savings account. The money draws interest and continues to grow. Each deposit increases the person's financial security in case of an unexpected need or emergency. Community equity works the same way. Each time the fire department meets the needs of a customer, a deposit is made into the community account. If the department is active throughout the entire community, and is providing quality service, this account will grow quickly. Then, when the organization needs the support of the public, the support will be available.

For example, consider a department that has developed a large amount of community equity. The department needs to develop a neighborhood vaccination program, but requires an increase in the budget. The community equity can be used to gain support for the budget increase.

A word of caution! Just like a real savings account, the fire chief can't take out more than is put in. Similarly, if the fire chief continually makes withdrawals from the department's community equity "accounts," the balance will eventually be zero. Community equity is best saved for the "big" issues such as improving services, improving working conditions for internal customers, and developing department infrastructure.

Provide Services at the Neighborhood Level

In many cases in the fire service today, services are provided as a "whole-piece" throughout the community. The services and programs delivered in one area are the same as those in other areas. In the future, services and programs will be based upon the specific needs of a neighborhood. The days of "one size fits all" will no longer apply.

This new paradigm is being driven by the growing diversity found in communities. Many U.S. towns, cities, and rural areas are a mix of people with different cultures, income levels, and expectations. These groups have different needs. Fire and injury problems may vary from one neighborhood to another. The local fire department will have to develop and implement services and programs that will meet these varying needs.

To find out what these needs are, fire department personnel must interact continually with members of each neighborhood or community. In a sense that is "back to the future," back to the days when the fire station was an important center of activity in the neighborhood or community. Community activities and meetings were held at the fire station. Residents knew that the fire fighters could help when they had a problem. Today and in the future, outreach involves not only seeking input from the public about needs and programs, but also having fire fighters providing PIER programs to the public. This is especially true for fire safety and injury prevention education. For some departments, this will be done during duty hours on a regular basis. For volunteer departments, this may be accomplished during specific times in the year or on a weekend. The days of only existing to respond to an emergency are over. In the next century the non-emergency role of firefighters will be a critical element in the success of any fire service organization.

"Reaching out to the people in each individual neighborhood must be a priority for every fire fighter in today's fire department. Neighborhood outreach is the only way we have to provide the quality of customer service our customers, our owners, expect and deserve! You simply can't provide customer service from a recliner in a station."
— Frank Carter, Deputy Chief, Colorado Springs, Colorado, Fire Department.[9]

Finally, neighborhood-based programs are the best tools for developing community equity. The fire department must become a vital member of the community, as much so as any other government agency, private business, or nonprofit organization.

Involve Customers and Employees in Decisionmaking

"Odds are good that the best ideas will come from people who both understand the goals of the organization and are working directly with the communities they seek to serve. This requires involving them in substantive decision making."
— Meri-K Appy, Vice President, Public Education, National Fire Protection Association.[5]

Tom Peters, well-known management guru, preaches that the best organizations are those that seek input from employees and allow them to assist in the decisionmaking process. This means more than simply asking a few questions about current issues. It means reaching out to every part of the organization—operations, fire prevention, community education, and training—and getting the people involved in solving problems and in developing the organization's future. It is the men and women who respond to the fires, provide emergency medical care, and teach children fire safety who really understand the needs and capabilities of the department. They can provide the best information on what is needed where, and usually the best way to get there.

This is not to say that fire departments don't need leaders. They certainly do! However, the most effective leaders are those who use the best resources available. In this case, the resource is people!

External customers will also be active in providing input in the decisionmaking process. The citizens of the community are in a real sense the "owners" of the business. Their taxes pay for the services and programs of the fire department. It makes sense to get them involved when possible. This approach also taps into the great wealth of management experience available in a community. It is yet another way to increase the department's community equity, and it improves the decisions made by the fire department leadership.

In the future, fire service leaders will need to develop internal and external processes for seeking input on issues, problems, and quality of service. Internal teams will have the authority to develop and implement new programs. New ideas will come from all parts of the organization. Community focus groups and committees will encourage citizen involvement in key issues. Community meetings and the media will be used to get feedback on the quality of services being provided. The new paradigm will require innovative approaches to gaining and keeping employee and citizen involvement in decisionmaking.

Focus on Prevention and Education

"Prevention education, more than anything else, seems to work."
— Leadership in Public Fire Safety Education: The Year 2000 and Beyond.[4]

As financial resources available to the fire service become more difficult to obtain, the need to implement the most effective and successful programs will become greater. Of course, success will be measured in terms of "preventing harm" to the customers: reducing the number of fires; reducing preventable injuries; increasing the number of installed, working smoke detectors in the community; and the like. The best method for achieving this success is targeted prevention and education programs.

However, there is more to prevention than public education. Comprehensive prevention programs are made up of three components known as the three Es: education, enforcement, and engineering. Public education, as discussed in Chapter 1, involves changing the public's attitudes and behaviors so that people lead a safer life. Enforcement involves adopting and enforcing building codes to make the community safer. Engineering involves designing and building structures so that they are inherently safer. Engineering also involves the use of fire protection systems such as fire detection alarms and sprinklers to increase the level of protection in a building. The most effective prevention programs use all three components to increase the safety of the individual citizen and the overall safety of the community.

While these types of programs are not new to the fire service, they will see a dramatic evolution in the future. Overall prevention efforts will focus on specific problems rather than taking a "shotgun" approach. In other words, the greatest problems, whether fire or injury, will receive the greatest attention.

The scope of potential prevention issues is broad, and will include such things as wildland fire mitigation, firearm safety, and wellness. The future will see more and more strategic partnerships used to develop and deliver community education programs. New types of fire protection systems will also be developed more quickly.

Another area of prevention that will have a prominent role in tomorrow's fire service is managed health care. Managed health care, in simple terms, embodies the philosophy of prevention. Managed care seeks to reduce costs of health care by preventing injury and illness whenever possible, and then controlling the costs of care. Fire departments will be actively involved in the managed care system in each community. Much of this involvement will be in injury prevention and wellness education at the neighborhood level.

The fire service of the future will need to become the community leader in preventing harm through targeted prevention and education. If not, this role will be assumed by other government and nonprofit agencies, and the fire service will risk losing precious resources and community support.

> *"The biggest change that is looming on the horizon appears to be that of managed health care. The basic direction of health care over the years has been one of consumption. Managed health care changes all of that and focuses on conservation."*
> — *Dennis Rubin, Chief, Dothan, Alabama, Fire Department.*[6]

Create High Return on Investment

Investment. High return. These are terms used more frequently in business than in the fire service. Yet, in the future, the citizens of each community, the "owners" of the local fire department, are going to expect the department to provide a high return on their investment—their tax dollars or the donations and financial support to volunteer fire departments. More than ever before, customers are going to expect more from their fire department than simply putting out fires once in a while. They will often expect the fire department to provide new programs and services with the resources at hand.

This customer expectation will be met in large part through careful budgeting and quality management of current programs. Fire chiefs will need to evaluate each and every operation in the department and ensure that each makes the best use of appropriated funds. Where there is inefficiency or duplication, changes will have to be made to make better use of resources.

This expectation will also be met by using current resources to implement new programs. Initially, this may not seem possible. This approach to resource utilization is called "value-added" service, and is a concept that has been practiced in business for more than a decade. Value-added simply means getting more use out of current resources for the benefit of the customer.

A good example of a value-added service is the use of operational companies to conduct home fire safety inspections. The customer is already paying for these companies to be in the station ready to respond to an emergency. In a value-added approach, the companies are used to conduct home fire safety inspections when not involved in emergency operations. This philosophy of resource utilization is ideally suited for fire and injury prevention programs, neighborhood wellness programs, and programs designed to market the department's programs and services.

A good example of value-added service for volunteer fire departments is a program which makes available chimney brushes for use by the public. In many rural areas of the country, chimney and flue fires are a problem. By making available chimney brushes to the public at minimal or no cost, the department is helping improve the fire safety of the community, and is making available a "service" that might not otherwise be available or affordable to the citizens.

Another way to build value-added service is to use support resources to generate revenue and/or to offset costs. For example, there are many mandated hazardous materials training regimes required of industry. A fire department also provides the same training to its personnel. Why not provide the training to local industry for a fee? Why not use apparatus maintenance facilities to contract maintenance service to smaller departments? Why not provide first aid and CPR training to the public for a small fee?

Partnerships that pool resources will be formed, resulting in a program or service that otherwise would not have been possible. For example, a partnership between a local community college and a fire department for delivery of fire training saves money and expands the availability of training.

All these types of programs increase the yield of the department's budget. This increase will ensure that the customers' expectations for effective use of tax dollars are being met.

> *"The competition between governmental departments is fierce, and should not be underestimated. By meeting or exceeding the customer's expectations and ensuring that there is added value to our service, the citizens will be much more generous with financial resources."*
> — Dennis Rubin, Chief, Dothan, Alabama, Fire Department.[6]

"A paralyzed senior management often comes from having too many managers and not enough leaders."
— *John. P. Kotter, Leading Change: Why Transformation Efforts Fail.*[10]

Leadership is something much different from effective management. Managers work with things: facts, figures, apparatus, equipment. Leaders work with people: the men and women of the fire service. The fire service is going to need leaders who can pick up the banner and lead the charge into the next century. There will be a new focus and drive toward fostering this leadership into a strong force that can create a vision for the future, and then bring it into reality. For volunteer departments, a key leadership issue in the future will be the recruitment and retention of volunteer firefighters. Many departments are struggling already to maintain membership levels. The future leader in volunteer organizations will have to be able to overcome this challenge.

These new leaders will form new alliances with members, treating them as partners in the business. They will bring new ideas and programs to fruition. They will aggressively tell the story of the fire service—marketing the programs and services to the customers. They will mentor other leaders from within the department.

These leaders will not only be the fire chiefs; they will come from all levels of the organization—firefighters, operators, inspectors, company officers, paramedics, public education specialists, secretaries. Leadership will not be confined to the few with crossed bugles and gold on a badge. Leaders will be dedicated to the mission and the vision for the future; their focus will be on people.

"The paradigm shift taking place today is from management to leadership. Leaders create an environment where employees make the day-to-day tactical decisions. Leaders provide the tools and the training the employees need and then empower them to do it. Leaders 'get out of the way.' Instead of control, they give up control. As you give up control of what you thought was important, you gain control of what is really important."
— *Howard E. Hyden, Chief Executive Officer of Hyden & Hyden.*[11]

Emerging Issues Will Alter Paradigms

Before fire department leaders can begin the change process, they must understand the issues a department faces. While each department may face different local issues, all will have to grapple with certain national issues.

These national issues have been identified in an ongoing process, the Wingspread Conference.

The purpose of the Wingspread Conference—first begun in 1966 and held every 10 years—is to examine current and future issues facing the fire service. The Fourth Wingspread conference was held October 23 to 25, 1996, in Dothan, Alabama. At the meeting were some of the most prominent leaders in the fire service, including Chief Alan Brunacini of the Phoenix Fire Department, Dr. Denis Onieal of the National Fire Academy, Rich Duffy of the International Association of Fire Fighters (IAFF), Nancy Trench from Oklahoma State University, and Chief Dennis Rubin of the Dothan Fire Department.

The participants identified both ongoing and emerging issues of national importance to the fire service. Those issues are summarized below. (Also see Appendix A, "Wingspread IV, Statements of Critical Issues to The Fire and Emergency Services in the United States.")

Emerging Issues of National Importance

Customer Service: The fire service must change its focus from the traditional emphasis on suppression to a focus on discovering and meeting the needs of its customers.

Managed Care: Managed care may reduce or control health costs. It will also have a profound impact on the delivery of emergency medical services.

Competition and Marketing: In order to survive, the fire service must market itself and the services it provides, showing its customers the value of what it does.

Service Delivery: The fire service must have a universally applicable standard that defines the functional organization, operation, deployment, and evaluation of public fire protection and emergency medical services.

Wellness: The fire service must develop holistic wellness programs to ensure that fire fighters are physically, mentally, and emotionally healthy and receive the support they need to remain healthy.

Political Realities: Fire service organizations operate in local political arenas. Good labor/management relations are crucial to ensuring that fire departments have maximum impact on decisions that affect their future.

Ongoing Issues of National Importance

Leadership: To move successfully into the future, the fire service needs leaders capable of developing and managing their organizations in dramatically changed environments.

Prevention and Public Education: Fire and emergency services must continue to expand the resources allocated to prevention and education activities, the goals of which are reducing injuries from fire and other risks.

Training and Education: Fire and emergency services managers must increase their professional standing in order to be on an equal footing with their peers in other disciplines. This professionalism should be firmly grounded in an integrated system of nationally recognized and/or certified educational training.

Fire and Life Safety Systems: The fire service must support adoption of codes and standards that mandate the use of detection, alarm, and automatic fire sprinklers, especially in residential properties.

Strategic Partnerships: The fire service must reach out to others to expand the circle of support to assure reaching the goals of public fire protection and other emergency and prevention activities.

Data: To successfully measure service delivery and achievement of goals, the fire and emergency services must have relevant data and should support and participate in the revised National Fire Incident Reporting System (NFIRS).

Environmental Issues: Fire and emergency services must comply with the same struggling Federal, State, and local ordinances that apply to general industry and which regulate response to, and mitigation of, personnel safety, and training activities relating to the environment.

The Change Process

Both the fire service paradigm shifts and the issues identified at the Wingspread Conference point to the same thing: to thrive in the future the fire service must change. For some departments, the changes required will be minimal. In other departments, simply surviving will require a wholesale change in ways of thinking and operation. Whether the change will be small or large, it will begin in the same way: through the breaking of old paradigms and the establishment of new paradigms. This is the change process.

Change, for most people, is frightening. People like to have at least some control over what happens in their lives. Organizational change generally represents a loss of that control. In the past employees have had little, if any, say in the direction of change. Yet it doesn't have to be that way in the future.

Change that is unforeseen and wreaks pain within an organization is usually change that is imposed from external pressures. In these cases, management of the organization has not had the foresight to recognize the need for change, and then must take drastic steps to deal with the changing environment.

A good example is a well-known corporation which, for decades, was the world leader in computing. No other corporation could match its breadth of products and services. To management everything was fine, and no change was needed. After all, the programs, services, and organizational structures had worked well in the past, so why fix it if it isn't broken?

In the 1980s, the computer world changed dramatically. Technological advances occurred very quickly. There were changes occurring in the work force, society, and in the competition. The corporation's management knew that they needed to make changes to survive, but by the time they acted, it was too late. The world leader in computing had fallen behind the competition. The corporation had no choice but to take drastic, painful measures. Parts of the company were sold. Employees were laid off for the first time. Salaries were cut. In the end, the unwillingness or inability to drive change internally resulted in the external environment driving the changes. This scenario, almost without exception, ends in turmoil and loss to all involved: employees, customers, and management.

But change doesn't have to occur that way. Change can be directed. The need for change can be recognized early and the whole process implemented with positive results. The key to directing positive change lies in recognizing the need to change early, and involving of all the stakeholders in the change process.

Hewlett-Packard is a good example of a corporation that successfully directs change. Hewlett-Packard, a technology company similar to the corporation in the previous example, has a history of successful change. The people at Hewlett-Packard, from the top to the bottom, are always looking for ways to do things better, looking for new products to meet the current and future needs of customers, and keeping a close eye on what's happening in the world. As a result of this proactive approach to change, Hewlett-Packard many times has created the change that the rest of the industry has followed. Change has become part of the way of doing things at Hewlett-Packard.

The same has been and will be true in the fire service. Those leaders and departments that recognize the need for change, and then pursue that change in a planned and deliberate process, will reap the rewards.

During change in an organization, there are four elements as shown in Figure 2, "Elements in the Organization Affected by Change," that are most affected by the change process:

- services and programs,

- organizational structure,

- training of employees, and

- values and paradigms of the organization.

Change will almost always result in modified services and programs. In the fire service, one thing this may mean is a shift from the strong focus on response to a strong focus on preventing harm before it occurs. To accomplish this change in services and programs, there will certainly need to be changes in the structure of the organization. These new services and programs are going to require training of employees. Finally, the changes will be rooted in the breaking of old values and paradigms and the creation of new ones.

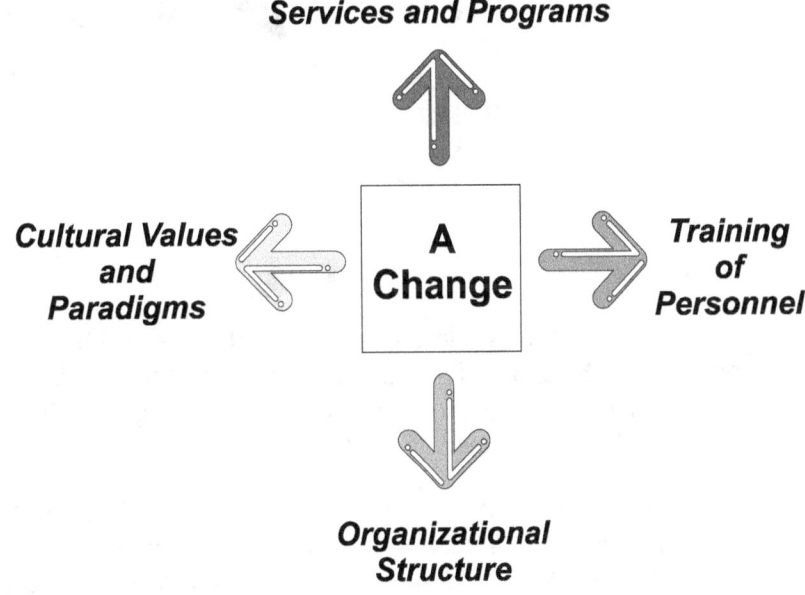

Figure 2. Elements in the Organization Affected by Change.

Change, to be successful, must involve all the stakeholders affected by the potential changes. For the fire service, this includes:

- Internal customers,

- External customers,

- Department leaders,

- Elected officials,

29

- Local business, and

- Local government.

See Figure 3, "Stakeholders in the Change Process."

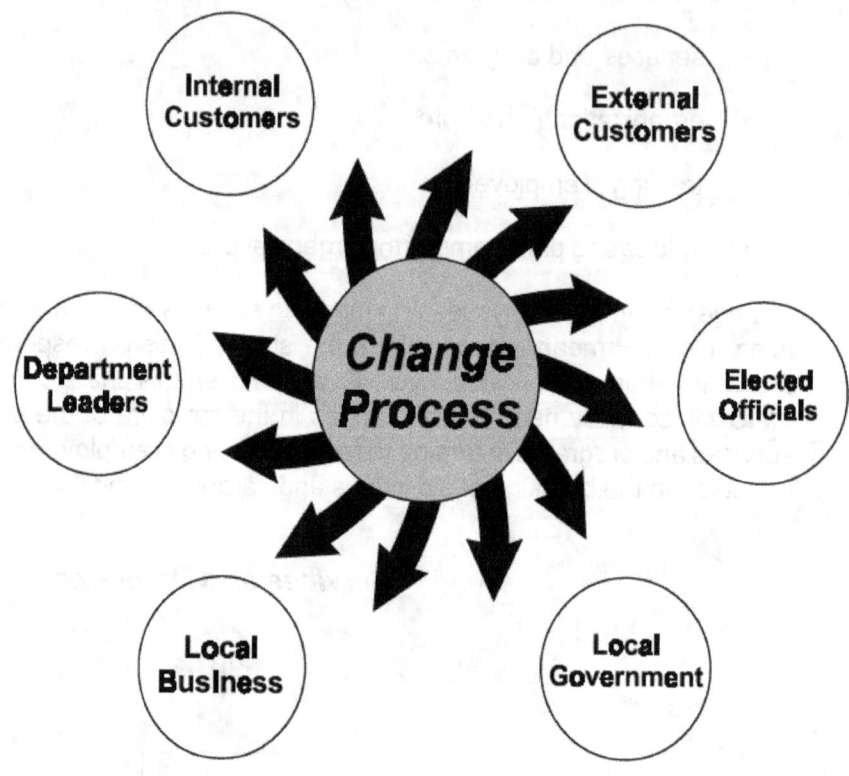

Figure 3. Stakeholders in the Change Process.

The internal customers, the employees, in the end will be the ones who must make the change work. Keep in mind that involvement doesn't necessarily mean they will like the changes, or in the beginning see the need for change. But their involvement will help ensure the ultimate success of the change process.

> *"Top management needs to talk less and listen more, to trust and depend on others in ways they never have before, because the answers to our problems will come from those who are closest to those problems-and closest to our customers."*
> —Ken Blanchard, Author of The One Minute Manager.[12]

The change process must also involve the external customers. Their needs will be one of the strongest driving forces for change. Their involvement begins with the leaders in the department establishing relationships with the different customer groups and asking them about their needs, now and in the future. This is not a one-time event. Communication with external customers, as well as with all other stakeholders, must be ongoing and must be something to which department leaders are truly committed.

If this communication is ongoing, taken seriously, and used effectively, then the department will be in an excellent position to deal with a critical stakeholder group—the local elected and appointed leadership. This stakeholder group is extremely important because it ultimately makes the decisions about community resource allocation that affects service provision across the board. Remember that the influences driving change in the fire department also are driving change in every other municipal department . Furthermore, your external customer— the citizen—is the customer of the other municipal departments, too. The customer wants resources allocated where he/she perceives the greatest return on the investment, and the local appointed and elected leadership makes these resource allocation decisions on the citizens' behalf. You need to be prepared to show the local elected and appointed leadership how you are meeting customer needs.

> *"The local union is usually a potent political and social force in the community. We can and we should use that power to assist fire chiefs on issues that directly affect our members, and even on issues that don't. Assisting a fire chief in promoting policies to benefit the fire department and the community can go a long way in solidifying a positive relationship."*
> — Alfred K. Whitehead, General President of the International Association of Firefighters.

Steps in the Change Process

When initiating change from within or directing change that is occurring due to changes in the external environment, there are steps that can be followed that will help bring about the desired change. These steps cannot simply be taken on a whim. Rather, they require the organizational leadership to plan, prepare, and make a commitment to achieving success with the change, knowing that most successful change requires time, sometimes as long as years. The steps are illustrated in Figure 4, "Steps in the Change Process."

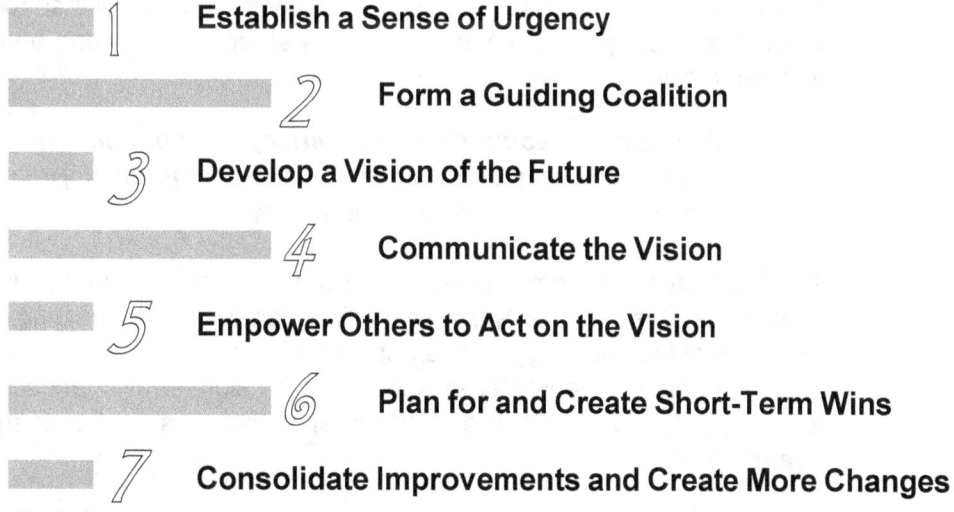

1 **Establish a Sense of Urgency**

2 **Form a Guiding Coalition**

3 **Develop a Vision of the Future**

4 **Communicate the Vision**

5 **Empower Others to Act on the Vision**

6 **Plan for and Create Short-Term Wins**

7 **Consolidate Improvements and Create More Changes**

Figure 4. Steps in the Change Process.

The first step in the change process is to establish a sense of urgency. As mentioned earlier, change involves breaking old paradigms and forming new ones. Most people won't break out of their old models of behavior unless they believe there is a compelling, urgent reason to do so. The leadership of the organization must communicate to everyone that there is danger ahead unless something changes. In short, if the organization keeps doing things the way it has in the past, the organization may not survive.

A good example of this in the fire service is the use of personal protective equipment during emergency medical incidents. When personnel began to understand the risk of exposure to infectious diseases while giving medical aid, there was an urgency about the need for protective equipment. Without the urgency and personal risk of continuing to do things the old way, there would not have been change.

The second step in the change process is the formation of a guiding coalition. The purpose of the coalition is to initiate and direct the change process. The coalition should involve leaders from the different stakeholder groups. These leaders must have a good sense of the needs of their groups and have the authority to influence the change. The department administration, the chief officers, must also be represented.

The coalition must be small enough to be able to conduct business effectively. The group must have the total support and endorsement of the chief executive officer, as a change effort will fail without support from the top.

The third step in the change process is to develop a vision of the future. This vision should be created by the chief executive officer, and then communicated to the guiding coalition. The vision makes the future tangible, something that can be seen and worked towards. A vision helps to motivate and direct the actions of the employees, and it is the blueprint that is used to develop the strategy for the change process. Without a clear, easy-to-understand vision, the desired changes may never materialize.

For example, the vision of "preventing harm" is nebulous and could mean many things to different people. However, the vision of having "neighborhood education programs that reduce preventable injuries to our children" is more tangible and gives both the guiding coalition and the other stakeholders something at which to aim their efforts.

Leaders take note! The vision must be clear. It must be something the stakeholders can believe in and work toward; it must create a better tomorrow than today; and it must be a vision that is easy to communicate. If it is so complex it can't be easily explained; it is doomed to fail.

> *"A group of people truly committed to a common vision is an awesome force. They can accomplish the seemingly impossible."*
> — Peter Senge, Author of The Fifth Discipline.[13]

The fourth step is to communicate the vision. If there is a step that is decisive in the quest for success, this is probably that step. The members of the guiding coalition and the chief executive officer must aggressively carry the banner of the vision. They must tell everyone about the vision and about every aspect of the vision—why change is needed, the benefits to change, the person's role in the change process, the consequences of not changing.

This communication must occur both formally and informally. The message must be communicated to the same people more than once, in more than one method. Publish it in the department newsletter. Hold a department meeting. Make a video.

As the process gets started, the vision should also be communicated by managers and supervisors. This communication and support should be a part of supervisors' and managers' jobs. The vision is about the future of the department. It is a priority for everyone in the department. Everyone deserves to hear the vision so that they understand all of the issues.

A final word about communicating the vision. The most prominent leaders in the department, especially the department chief, must be seen and heard frequently communicating the vision. The leaders must be available to answer questions about the vision. This will require open, honest discussions in face-to-face environments. There is no alternative to this investment in time if change is going to be successful.

Change in a department is something that can't be accomplished by a single person, even the fire chief. Most change has an impact on all areas of the department and on all the people. In fact, change has to do with people more than anything else in the department. For change to occur, the men and women tasked with making change happen must be given the authority, freedom, and resources to make it happen. They must be empowered to bring the vision to a reality. This is the fifth step in the change process.

This is not to say that they have a free rein. Certainly there is accountability. But the control so often hoarded by the chief executive officer has to be relinquished to those who are implementing the vision. Nothing destroys a change process faster than micromanagement.

Part of this empowerment is listening to feedback once the change process is underway. The change process is not an absolute. It must be modified and revised as time goes on. Without open communication between those directing the change and those carrying out the change, revising the course of the change is very difficult.

> *"Labor and management each have their strengths and each have their weaknesses. We can work together—as strong local unions allied with strong fire chiefs—and make great strides in managing future changes to benefit our members and our fire departments; or our local unions and fire chiefs can work separately—and let the changes manage them."*
> — *Alfred K. Whitehead, General President of the International Association of Firefighters.*

Another part of empowerment is allowing personnel to take risks that will improve the change. For example, the change may involve implementing a new home inspection program in a specific neighborhood. The inspections are conducted by on duty personnel. During the course of implementing the program the personnel find a better way to notify the public about the program. They should be able to take a risk and try the new method. If it doesn't work, that's fine. If it does, the change process is improved. There must not be a penalty for trying something new and innovative as long as it is not in violation of department policy or applicable laws.

> *"Letting people take the initiative in defining and solving problems means that management needs to learn to support rather than control."*
> — *Ronald Heifetz, Harvard University School of Government, and Donald Laurie, Director of Laurie International Limited.*[14]

Planning for and creating short-term wins is the next step in the process. Successful change involves taking many small steps rather than a few big ones. It's like the cliché about eating an elephant one bite at a time. The strategy for change should be based on an accumulation of small changes that are easier to achieve. This is especially important in organizations that have had bad experiences with earlier change efforts.

For example, if a vision for change is to ultimately create a physical fitness and wellness program, a small step could be the purchase of fitness equipment for a station and allowing firefighters to use the equipment during on-duty hours. Each individual step allows the personnel to accept the change, to "buy into" the use of the training equipment. Eventually, all the steps taken together will result in both the creation of a fitness program and acceptance by firefighters of the new program.

Leaders must not leave these short-term successes to chance. They must be incorporated into the change strategy. Important steps left to chance will not happen during the change process.

Finally, change itself can be a new paradigm. The most successful fire service organizations accept change as something that is positive and as something that is necessary to move into the future. This acceptance is a result of many successful and positive experiences with change. Creating short-term wins provides those positive experiences to the employees, helping to remove the fear and resistance to change.

The final step in the change process is to consolidate improvements and create more change. In most organizations there will be several different changes occurring at the same time. As these changes are completed, and the benefits experienced, the leaders must communicate the success to all the stakeholders, both internal and external. The improvements must become part of the organization, part of the culture.

The new improvements should be used to foster even more changes, which will then result in more improvements and a higher quality of service to the customers. The need for change is ongoing. Departments never "arrive." They are constantly seeking to provide the best quality service, to implement new programs to better meet the needs of the customer, and to improve conditions for the internal customers. One successful change should be used as a springboard for even more changes. This process of change will help create an organization that is leading the future rather than reacting to it, an organization that becomes the model for tomorrow rather than its victim.

Competing in the Future

Change for the sake of change is not only a waste of time, it is not healthy for an organization and its people. The purpose of change, of looking to the future and changing to be ready for it, is to have a healthy fire service organization that is able to meet the needs of internal and external customers. In a word, the purpose of change for the fire service entering the 21st century is to thrive—to be able to compete for resources and community support so that the needs of the customers can be met.

The remaining chapters in this manual provide a road map for competing successfully, based upon the future paradigms of the fire service and society. Developing and implementing effective community education, information, and relations programs are emphasized.

PIER programs are essential to reaching out to the customers at the neighborhood level and providing services and programs. As discussed many times earlier, preventing harm through prevention and education with neighborhood programs is expected by the owner of the business, the customer, the taxpayer. In short, it is the future.

A closing thought about change and the future. The leaders in the organization have one final responsibility regarding change—to ensure that it becomes institutionalized as part of the organization's culture. This is done by developing and mentoring new leaders for tomorrow, preparing the next generation for leadership and change. It is done through creating a department structure that allows open communication. It is done by adopting policies that encourage employees to try new things, to experiment with solutions to problems. It is done by getting supervisors and managers involved with the change process so that it can be better managed in the future. It is done by fostering partnerships between management and labor with a goal of improving the service provided to the customer.

Finally, it is done by acting, by doing something, by getting up out of the chair and making it happen. Change, success, survival in the future—all require leaders to lead, to help people to see the vision of the future, and then to bring that vision to reality.

> *"So, take the initiative and act. Your basic nature is to act, not simply to be acted upon. By choosing your response, you can create your own future."*
> — *Stephen R. Covey, Author of Seven Habits of Highly Successful People.*[15]

CHAPTER 3

> *"I know you believe you think you understand what you do, but I'm not sure you realize that what you think you are doing is not what you should be doing today!"*
> — *Anonymous*

What was done yesterday may not provide solutions to tomorrow's problems. Albert Einstein had a quote on his office wall which simply stated "The significant problems cannot be solved at the same level we were at when they were created." It's a changing world, with changing problems that need changing solutions. Change is an attitude and a process.

What Do You (the Fire Department) Really Do?

Discovering what it is you really do is part of your journey to providing the best possible service you can. You must be willing to look honestly and deeply into the work you may love, but which, in all honesty, may not be providing the results that will truly make a difference in loss of life and property in your community. It is critical that you be willing to become a change agent and to retool what you do, if necessary. One way of doing this is through benchmarking.

Benchmarking is a process used by organizations to assess their own performance. It is a tool to help the department set goals for itself, and to focus on attaining them.

An example of benchmarking was the process many municipalities used in the mid-1980s to determine comparable worth to establish competitive pay ranges for various job classifications. Another example relating to the fire service, although largely immune from competition, is the fact that other services in communities have privatized over the years. The public is putting pressure on fire departments to be more cost effective. Benchmarking can support a top-performing department during periods of fiscal stress by providing fire managers, city administrators, and elected officials with fact-based feedback on their performance. It can also provide the impetus for poor-performing departments to examine their processes and implement improvement programs. Perhaps you don't need a "full paradigm shift, just a new drill bit!" But you won't know unless you look.

To start, there are several data and informational elements that need to be studied throughout the organization. First, review what services you provide to the community. Make a list. Seeing everything on paper gives you a new perspective of how broad the scope and the speciality of the services you provide to the community really are. Your list may include some of the following services.

Response Services: These may include suppression activities, EMS, hazardous materials, water rescue, high-angle rescue, confined-space rescue, search and rescue, assisting the police department, disaster recovery, fire investigation, incident data collection and analyze, training to ensure safe and effective response, and yes, even rescuing animals.

Prevention and Enforcement Services: These activities may include code development; inspection activities; plan review; meeting with building officials, developers, owners, and business operators; enforcement issues; preplanning; right-to-know issues; data collection on prevention activities; juvenile firesetter programs; public information activities; and other special enforcement and prevention activities that communities may elect to have the fire department perform, such as supplementing neighborhood watch programs, being visible in the schoolyard after school, painting hydrants, installing numbers on properties, and many other unlimited creative activities.

Public Education Services: May include teaching community members the fire safety behaviors necessary to stop a fire from happening or how to survive should one occur; working with juvenile, youth, and child firesetters and families; providing referrals to the legal system and mental health communities; being a resource or liaison to schools implementing a school-based fire safety curriculum or program; providing programs for the elderly and other high-risk populations; designing programs and materials based on data and information collected on community needs; and being a grant writer, solicitor of resources and funds, program evaluator, analyzer, and trend explorer.

Fire departments provide a host of valuable services to a community. When services are viewed in this manner, we get a better perspective of how many specialized functions are provided.

While a public educator may view public education as the most critical function in the organization, one must realize that all functions are important to serving a community. The major difference is that public education services are the one way to stop fires and injuries from happening in the first place.

The response and enforcement services assume there will be a fire (failure) that will have to be extinguished or that buildings must have built-in safety features to protect people when a fire does occur. Fires are a people issue! Consider that it is not a cigarette that falls asleep while being smoked, and it is not the cooking pot on the stove that decides to leave itself unattended. If we educate people to understand fire behavior and to be good decisionmakers, we will be able to better accomplish our mission of serving the public and protecting life and property.

One should not assume that if we are successful in preventing all fires, that we will lose jobs and resources in the fire service. Dr. William Hayden, an injury control specialist states "Prevention will not cut resources; it will in fact generate the need for better equipment, people, and resources, because of the injury control necessary after an incident occurs and the ongoing effort necessary to provide the public with education."

Public education's goal is to lessen the impact and to stop as many fires, injuries, and deaths as possible. Lack of knowledge results in carelessness and should not be tolerated by any community. It is time to stop our tolerance of fire. It destroys more than $8.5 billion dollars in property each year, kills nearly 4,500 people, and inflicts burn injuries on millions. The cost of tolerance and ignorance is high.

What Are Your Services Worth?

Define the services you provide in terms of quantitative and qualitative value. Suppression is one of the most costly services in the fire department, and when the alarm sounds, we need to be ready to respond. Suppression consumes, in most cases, the vast majority of fire department budgets, yet may make up only 7 percent to 10 percent of what the fire department does. While the cost is high for suppression, the value may be high as well if the community is unwilling to absorb the expense of losses that occur. However, if the loss is acceptable, then the value of suppression services to the community may be deemed less valuable.

One of the biggest mistakes we make in the fire service is the way we report fire losses. We report numbers of incidents, deaths, injuries, and dollar loss. The paradigm shift in reporting that we should consider is that of reporting the value of the property we save, the number of smoke detectors that alerted people—ultimately saving lives, and the value of property that did not burn. This is a significant change in the way we have been traditionally taught to think about what we do. What this will do is add value to the invisible services we provide that the public never sees or hears about.

By defining the value of saving a life, a commercial or residential property, the value of the medical services provided, and the value of the service provided in a disaster situation, the community receives a different view of the value of fire department services. It has been said for every person affected directly by a fire, several other people are affected indirectly, and this includes losses to the tax base of a community. Think for a minute about the effect on a community of a fire in a facility that employs many members of the community or in an irreplaceable property, such as an historic building.

In Minneapolis, Minnesota on Thanksgiving Day 1982, two 13-year olds set fire to construction debris located adjacent to a 19-story office building. The fire caused $125 million in damage. Housed in the building was a prestigious law firm whose offices were consumed. Located in the office were the firm's historic law library and records. The contents were destroyed, and irreplaceable documents, books, and records were lost forever. The building had no built-in sprinkler system.

By defining the value to the community of protecting lives and property, we open up the possibilities of defining the value to the community of public education efforts.

What is the Fire Problem and Whose Responsibility is it?

In order to provide the best possible public education effort in the most cost-effective manner, we need to define the real versus perceived problems facing the community. Using a shotgun approach will not work; you must aim at a specific targeted behavior, audience, or location.

> *"Information is like a brick, it only clutters your yard unless you use it to build something."*
> — Anonymous

Specific information regarding the fire problem can be obtained in several ways. First, every fire department should be actively involved in an incident data collection system. The National Fire Incident Reporting System (NFIRS) was developed for use on a local level with a system to send the information to the State and onto the national database located at the United States Fire Administration (USFA). The NFIRS program is managed through a system of State and metropolitan program managers who work with the National Fire Information Council (NFIC) under a cooperative agreement with USFA. Of the estimated 30,000 U.S. fire departments, 14,000 are participating in the effort to collect incident information for a national database. This represents close to 50 percent of U.S. fire departments. The National Fire Protection Association (NFPA) is another excellent source of data.

While it is critical to have the fire problem identified at the national level, it is just as critical for local problems to be identified at the local level. The NFIRS program is more than just a method for reporting fires. It provides users with information on fire causes, response times, staffing and equipment used, deaths and injuries, activation of smoke detectors and sprinkler performance, equipment involved in the ignition, information on demographics, high-risk time of day, victims and behaviors, and much more. All of these elements can be used to identify fire problems, training issues, budget justification, and educational objectives.

The city of Cleveland, Ohio had a problem with young children playing with fire. In just 6 months, nine children died as a result of fireplay. To combat the problem, the fire department brought its case to the community. The department teamed with the local American Red Cross Chapter; together they decided the best way to reach children and parents was to initiate the NFPA Learn Not to Burn™ preschool curriculum in every daycare and headstart program in the city.

The Red Cross provided the curriculum and the NFPA conducted an in-service training along with the fire department. The response was overwhelming, and over the next two years the partners went one step further and implemented the Learn Not to Burn™ program in grades K-3.

Had it not been for available data and understanding the needs of the community the fire department might not have been as successful at solving the problem. Since the time of the implementation of this education program, not one child has died as a result of fireplay. This is a significant success story of a community coming together to solve a targeted problem.

In addition to the availability of data, it is important for public safety educators, public information specialists, and community relations specialists to avail themselves of education and training opportunities. FEMA's Emergency Management Institute (EMI) and National Fire Academy (NFA) offer excellent curricula in these areas.

Data are one of the most important elements needed to determine fire safety educational objectives. If you don't know where you are, how will you know how or when you get to where you want to go?

> *"Christopher Columbus is the only person known who...when he started didn't know where he was going; when he got there, he didn't know where he was; when he got back he didn't know where he'd been; and he did it all on borrowed money. In his time, he was never recognized for his contributions and died a poor man."*
> — Anonymous

An uncharted course will not lead to innovative, nontraditional efforts to solve today's fire problems. We must plan our efforts to provide the most cost-effective means to educate the public.

Each educational objective you plan should be measurable and evaluated at regular intervals. Having a grip on the real fire problem allows you to make decisions about what program outcomes are expected and how they should be measured. Defining the fire problem is valuable information that can be used when seeking funding, resources, support, and partnerships, both internally and externally.

In addition to NFIRS, other data sources may include investigative reports, local hospitals or clinics, insurance companies, police departments, health department statistics, schools, juvenile courts, prosecutors, fire service organizations, and arson and public education groups such as the Safe Kids coalitions.

Once you identify the fire problem in your community, you will have the ability to systematically find solutions. Determine who is at risk, what they are doing to put themselves at risk, and how to address the issues. Establish priorities. In most cases, you will not be able to address all the issues identified at once, so you will have to tackle them one at a time or find alternative solutions such as forming partnerships to help accomplish your goals. This process will take time. The educational process, however, is not nearly as risky to your firefighters or members of your community.

Whether you come from the ranks or are a civilian fire safety educator, it is important for you to understand the critical role you play in saving lives and property. Your training, expertise, and readiness are the tools you use to fight fire.

Honesty and Owning Your Results

Once you identify your services, measured against the fire problem, you have to decide whether these services meet the needs for solving the identified fire problems. This is the time for the entire organization, and perhaps the community, to be involved in becoming change agents. Why is this important? If your services do not provide a system to solve the fire problems, you may be putting your community, firefighters, and fire department at risk. The issue is balancing risk versus services. You should not view this exercise as a tradeoff between response activities and public education activities. It is intended to provide holistic service to the community.

If your department is not willing to work toward solving specific fire problems, you should be aware that other agencies and organizations will. If you need to retool your efforts or add resources, now is the time. Giving and getting the best bang for the buck is in the best interest of the fire department and the community.

The fire department has an advantage over the competition in providing solutions for a community's fire problems: it has a firsthand knowledge of the dangers of fire and how fire behaves. This knowledge is invaluable when working with the public in an educational setting. For instance, most communities place children's safety as a top priority. Use this as a springboard to address children's fire safety issues.

One method of addressing this issue is through the use of a school-based fire safety curriculum. Teachers teach the fire safety messages, which can be repeated often, in a way that is fun for the kids and is at the developmentally correct level for the children. The added value that the fire service can provide which other organizations can't is in service training for the educators relating the content material from first-hand experience. The teachers, in turn, teach the students in an educationally sound environment. As a public educator, your role is to nurture the program and assist as requested by the teachers.

Collaborating and forming partnerships to help solve some fire issues may be a step you need to take to ensure that your community is afforded the best possible knowledge, education, and skills. For example, the St. Paul, Minnesota Fire Department formed a partnership with Red Cross volunteers to provide training to senior citizens living independently in the community. They jointly developed a slide program that could be presented at churches and community centers to teach seniors about fire safety. The Red Cross volunteers were trained by the department and they then took over the tasks of scheduling and presentations. This partnership was crucial to reaching a high-risk population. The fire department only had two public educators, and there were thousands of seniors, living independently in St. Paul who needed to be reached.

In addition, the department and the Red Cross developed a home inspection guide and a system to install smoke detectors if the senior was in need. Another partnership evolved based on discussion with the trainers that identified the need for maintenance and repairs of heating equipment in the senior community. Local heating contractors and labor unions joined forces, and during four weekends before each heating season, seniors in need have their heating systems inspected and repaired for free. Consequently, St. Paul has lost only one elderly member of the community to fire since these partnerships were formed.

There are success stories like this all over the country. The fire problem is a community problem, and it is the people affected who are the victims. Therefore, they should be part of the solution. By involving the community, the fire department can get the buy-in needed to support continuing efforts at every level of the organization.

> *"Cooperate and involve."*

> *"Remember the banana...every time it leaves the bunch it gets skinned."*
> — Anonymous

Other value-added services the fire department can provide involve people resources. Few organizations in a community can muster person power like the fire service in the event people are needed. The fire service responds to almost every type of emergency imaginable. Most firefighters are trained to provide emergency response services, are on call 24 hours a day, and have a mission to protect life and property.

If your department is a high-visibility fire department, it will survive because the community will know exactly what it is you do for them and how well you do it. They will champion your need for funding and resources.

> *"Money never started an idea...it's the idea that starts the money!"*
> — Anonymous

When you provide and justify the plan, have your data in order, and have the community involved in your ideas, communities are more likely to generate the resources you need to operate.

In the fire service, there is a tendency to dictate the needs of the community for fire protection services. We must include — not exclude — those we serve. We may also find that people who are on the fringes of knowledge about the fire service may very well ask the right questions that will lead us to new solutions. In fact, the experience of members of a community in their areas of expertise may provide the fire service with some very cost-effective and unique solutions to problems.

By keeping your community involved, you will be marketing your services and guaranteeing a better chance of survival and support, and of the community generating the resources for you to continue your services.

> *"If at first you do succeed, try to hide your astonishment!"*
> — Anonymous

CHAPTER 4   *I Think I Know Who They Are, I'm Not Sure I Know What They Want*

Demographics (Who Are They?)

When we talk about demographic information, some people have absolutely no idea of what it is or where to look. Finding demographic information is no different than finding information on the fire problem, and in some cases, it's easier. Most State and local planning agencies, as well as libraries, universities, and colleges, can provide demographic information. It is generally not hard to find sources for identifying information that will assist you in planning and developing your approach to solving specific fire problems in your community or in understanding the members of your community and who they are. This information will also assist you in establishing goals and objectives for your public education and public relations projects.

Many fire departments make the mistake of dismissing this step of the process for gaining an understanding of community. In fact, they will miss identifying valuable information that can generate previously unknown resources.

> *"Even if you are on the right track, you'll get run over if you just sit there."*
> — *Anonymous*

Demographics are the essential characteristics necessary to understand the high- risk audiences of your community before you begin to plan for services, programs, or public relations efforts. Demographic information can include a wide range of informational elements, including ethnic background in a particular geographic area, education level, income level, type and age of structures or housing stock in an area, whether residential property is rental or owner occupied, age of occupants, number of children per household, and what languages are spoken by members of a community. In smaller communities, demographic information will be very helpful and may not be difficult to research. In larger communities or large rural areas, it may be more time consuming to research, but it is essential information.

Demographic information, coupled with incident information, will provide a tremendous amount of knowledge about your target audience and, more importantly, about members of your community and how best to reach and/or approach them. It can be used for planning education programs, public relations efforts, strategic planning, grant writing, fundraising, policy development, and legislative initiatives.

An example of the use of demographics comes from Minnesota. In 1994, the State Fire Marshal proposed legislation to require smoke detectors in every dwelling with a sleeping room. Before this was even proposed, the data were gathered on fire deaths in residential occupancies and demographic information relating to housing.

The trend information showed that an average of 79 percent of all fire deaths occurred in residential property. Further, from 1989 to 1991, on average, in 70 percent of these fatalities there were no smoke detectors present or there were nonworking smoke detectors at the time of the fire. This incident information, reported by participating fire departments that protect 96 percent of the State's population, came from the Minnesota NFIRS system.

From the demographic information collected, the Fire Marshal was able to calculate the number of households that did not have working smoke detectors, based on the finding that 80 percent of homes had smoke detectors, yet at any one time, 50 percent were not working. The number of owner-occupied homes identified in the State totalled 1.2 million. These were not covered by laws initiated in 1980 that required smoke detectors to be installed in rental occupancies by the building owners. This meant that more than 310,000 homes were at risk. Further, 89,000 of these households were identified as having children or elderly with an average income below the poverty level.

This report was presented to a Governor's advisory council, consisting of 15 appointed member organizations with an interest in fire and life safety issues in the State. This diverse group of organizations, including building owners, architects, League of City members, fire service organizations, the medical community, and insurers, supported the proposed legislation. This presented a very united front during the legislative process.

Armed with this information and support, and a plan to reduce residences without detectors or with nonworking smoke detectors to 25 percent by the year 2000, supporters got the legislation passed. It became effective in August 1994. This provided a win-win situation.

Citizen awareness of the problem increased dramatically and, with the aid of smoke detector giveaway programs which cropped up all over the State, fire fatalities decreased. Partnerships were formed at the local and State levels to assist the fire service in reaching the 25-percent goal. Deaths due to missing or nonworking smoke detectors are down 38 percent, and the State is well on its way to its goal of getting a working smoke detector in 75 percent of all residences.

Demographic information can be significant for many functions in the fire service. It can be used to identify and target recruitment of firefighters for career, volunteer, or paid-on-call fire departments; to determine fire station locations; to determine the best location for special-function equipment such as hazardous materials units; to determine where to place protection for wildfire areas; to determine fire rates for certain locations in a community; and to justify services and resources for the fire department.

Another benefit of demographic information, touched on briefly in the beginning of this chapter, is its use in identifying groups with specific needs within the community. Over the past 20 to 30 years, many communities have seen dramatic changes in the makeup of their citizenry. For example, Roosevelt High School in Minneapolis, in 1964, had little diversity. Today, 43 languages are spoken at the school. The schools in the city of Boston require that every piece of literature sent home with schoolchildren to be available in 28 languages. In many small farming communities in the Midwest and upper Midwest, Hispanic migrant workers who came to work in the area decided to stay because of the quality of life. In most cases, the cultural differences are significant.

A fire department that identifies a high-risk fire behavior in a segment of the community or a geographic area will have to conduct research to understand and determine the methodology with which to reach the target population.

Demographic research can be complex or simple. If you need to use more complex research related to demographics, look for assistance from community members, libraries, city human resource departments, grant writers, colleges, or universities. Many colleges and universities offer course credits for research projects that can be done by interns.

As a fire service leader, it is important for you to understand how and when to use demographic information. The world today is information rich; we must learn how to use it to our advantage in reaching for our mission.

> *"There was more information produced between 1965 and 1995 than during the 5,000 year period from 3000 B.C. to 1965. The amount of available information is now doubling every five years—information is critical to existence today."*
> — *Anonymous*

What Does the Community Want?

Now that you have looked at incident and demographic information, you must not forget another level of information that is critical to the efforts and survival of your fire department in the future. You need to identify what the community perceives as your function and service and, further, to determine what they think they need or want.

The results may, in some cases, be a surprise to you if you have not done a community survey in the past. One fire service leader who recently retired stated, "It was a great surprise to discover that the community did not revolve around the fire department. In fact, little was ever reported in the media and very little community involvement was ever asked for." We are sometimes so close to our organization that it is easy to get the false sense that the community knows and is involved in everything we do.

Discovery of what the community knows or wants can take many forms, from a survey through the mail or on the telephone to focus groups, customer surveys, or simply setting up a system through other city services to ask decisionmakers, permission givers, and community leaders their opinions. However, you must be willing to accept the results and be willing to make some tough decisions about how you operate based on the information you receive and the needs expressed by the community. In some cases, it may be that you need to reeducate the community. But remember, people will want proof of your claims.

> *"You must get close—intimately close—to your customers. Seek regular, direct contact with them. Build a strong relationship. Deliver the highest quality service possible, listen to their needs and develop a reputation for responsiveness. In the final analysis, customers (community) are your only source of job security. See your department as a service center."*
> — *Price Pritchett, Author and Publisher of Multiple Management and Organizational Change Books.*[16]

Your department needs to be businesslike, so you must treat this process much the same way a business would. Your goal is to satisfy your customers' needs, and you do this by finding out what they want. It is an important task. Just as it is critically important to identify the real versus perceived fire problems, you must understand your community and listen to their view of your services and what they need or think they need from you.

The organization that does not choose to look past the present, even if the present looks great, is going to be passed by. One way to determine needed changes is by staying close to the customer. This can be done in many ways. Also, realizing that customer values are dynamic is important. Determining customer satisfaction, therefore, is an ongoing process.

Community/Customer Surveys

Conducting customer surveys is one of the most effective ways of finding out if your organization is meeting your customers' needs and expectations. Customer surveys are a way to quantify the attitudes of your population.

> *"We were given two ears and one mouth...so you should listen twice as much as you talk."*
> — Anonymous

Conducting a survey may require resources outside of the fire department. It is important when phrasing questions to make them simple, easily understood, easy to analyze and measure, and non biased. You may not have the skills or resources to do this in-house. Again, colleges and universities or companies that specialize in this type of activity may be a good resource.

An important consideration when designing a survey is to make sure the information can be collected in such a way that it is easily assembled and analyzed. Part of the survey design is selecting or developing an appropriate method or instrument for collecting information. Questionnaires are a widely used survey instrument and allow collection of data from a larger number of people. If you are designing your own questionnaire, phrase questions so that answers can be given on a scale, such as strongly agree, agree, don't know, disagree, strongly disagree. Questions that require an open-ended response are more difficult to collate and analyze from a large number of people. You may also wish to consider using a software program that can organize the information for you.

To make the survey more effective, interview a few customers before writing the questionnaire. This will be extremely helpful in clarifying topics for the survey. The questionnaire needs to be short and easy to complete. One- or two-word responses or a check-off of correct answers should be designed into the form. Providing an incentive for completing the questionnaire is often helpful. Incentives may include a battery for a smoke detector, or a free home-safety survey. Incentives are endless and only limited by your own imagination.

(See Appendix B for examples of Phoenix, Arizona, Fire Department Customer Surveys.)

While conducting the customer survey, it is also a good time to ask questions relating to the values of your community. Why? Your services should mirror the values of the community in everything you do. If you do not have a very good idea of the community's values, you may be wrong in your approach to the service you provide and the way you interact with the community. It is better to ride in the direction the horse is going, and riding in the same direction as the community is critical.

> *"It is what you learn after you know it all that is of most value."*
> — Anonymous

We must retool ourselves and our organizations constantly or we risk becoming obsolete. The more you understand the community, the more valuable you become. Using surveys and determining customers' needs will provide you with a better package of knowledge and skills than the competition. If you do not take care of the customer, someone else will step up to fill the gap. Only you can make the difference.

Another way to determine customers' needs is to bring together specific members of the community and work directly with them to achieve a high degree of fire and life safety for the community. The partnership revolves around a sharing of ideas and concerns.

The Phoenix Fire Department (1,153 sworn personnel, 268 civilian personnel) Urban Services Division, formerly known as Fire Prevention, accomplishes this through its Community Partnerships Program.

Community Partnership seminars provide a means of sharing mutual concerns for the life safety of the community. The community has resources that can improve emergency operations; these resources are shared by the participants during these seminars. Educational partnerships and pre-emergency planning programs offer opportunities to share knowledge, skills, and experiences and achieve a common goal of enhanced life and property safety. Recognizing the importance of each others' contributions and needs is what makes this a successful program. The Urban Services Division encourages the participation and collaboration of community leaders, outside agencies, and other divisions of the fire department in developing strategies to meet or exceed the needs of their customers — the citizens of the City of Phoenix.

Being proactive, not reactive, is an excellent guideline for any department interested in customer service. Being able to anticipate a problem or issue before it happens is a way to keep your customers loyal and pleased with the services they are receiving. Listening is most important. Without knowing the perceptions of your customers you are unable to address issues or adjust your service levels before a problem arises. Communication means listening and asking key questions that will let you know the needs and wants of your customers.

Having a forum, like the Community Partnerships, where everyone has a chance to listen to each other is a win-win situation for all involved.

In another example, the Indianapolis Metropolitan Professional Firefighters Local 416, the Indianapolis Fire Department, and the Marion County, Indiana, Survive Alive Program teamed up in 1995 to present a powerful and comprehensive fire and life safety program to more than 20,000 children and adults annually.

Besides its important message, what makes the Survive Alive Program stand out is the method in which labor and management have come together for the citizens they serve.

The not-for-profit Survive Alive Program is located in a restored fire station owned and operated by Indianapolis Local 416 as its union hall. The facility is used by the Survive Alive Program for training and safety education. The assigned staff for the program are on-duty Indianapolis fire officers, and, in addition, all Indianapolis Fire Department suppression personnel and administration staff are required to perform one tour of duty each year at Survive Alive.

The joint program also provides its public educators with loans of curriculum; education materials; and firefighter turnout gear, and trains teachers; community workers; and other groups throughout Marion County in fire and life safety.

Union members from throughout Marion County participate in the program which benefits schools throughout the county.

"We have learned that a strong partnership with fire department manage ment on projects that benefit the community are a win-win situation for everyone involved: the fire department, the local firefighters union, and the citizens we serve."
— Tom Hanify, President of Indianapolis Metropolitan Professional Firefighters, IAFF Local 416.

Another example of a very successful Community Partnership Program is found at Children's Village of Washington County, Maryland. Children's Village is a comprehensive life safety education complex located in Hagerstown, Maryland which has a population of 40,000. Washington County supports a population of 121,000.

Children's Village is a fully accessible $1.5 million facility, constructed by and for the citizens of the community. It serves as a central location where life safety instruction is presented to all second graders attending Washington County public and private schools.

In 1986, the local telephone company teamed with the members of the Maryland State Police to discuss the concept of developing a safety education center. A steering committee was formed consisting of members representing corporate businesses, local government, interested private citizens, and all emergency service agencies in the area.

The local Board of Education approved the concept and provided the five-acre parcel of land where Children's Village stands today. Insurance coverage and transportation for all students is also provided by the Board of Education. Emergency services instructors received training from classroom teachers on age-appropriate lessons for the second- grade students who would be visiting the Village.

A local contractor approached the committee and asked to become the volunteer construction coordinator for the project. This individual then proceeded to encourage building supply and contracting colleagues to donate building materials to this worthy project. In-kind donations supported the majority of the construction efforts. This overall support from the community continues, as efforts to enhance the development and delivery of new and innovative life safety education.

Children's Village, in addition to delivering life safety education to second graders, is used for many purposes. The Hagerstown Fire Department uses the facility for juvenile firesetting intervention. Life safety continuing education programs are presented during the evenings and weekend hours to all age groups. During the summers, 5,000 people attend a life safety open house.

A citizens Board of Directors manages Children's Village as a tax-exempt, nonprofit organization. The Board meets monthly to govern the development, operation, and maintenance of the facility. The Board is comprised of local emergency managers, elected officials, and citizens of the community.

Children's Village is a wonderful example of a community's commitment to create an injury control center where quality, age-appropriate life safety education is provided and which leads to behavioral changes that reduce death, injury, and property loss.

Focus Groups

A focus group is a qualitative research methodology and a semistructured information-gathering method. The information that you gather is not necessarily representative of your customers as a whole; however, it does give you insight into the views and opinions of the particular customer group with whom you are interacting. Focus group participants do provide you ideas and impressions about your department. These ideas and impressions are valuable in helping you to get a picture of your department as the customer sees it.

The more successful focus groups usually consist of no more than 7 to 12 participants. Too many people in a focus group make it difficult to capture all the comments that are provided. Participants work with a facilitator/moderator to express opinions, and they discuss a specific topic. Often, the moderator is brought in from outside the department, and is usually someone who is very skilled in leading these discussions. The moderator needs to be very clear as to the goals of the session.

The objective of these groups is to acquire a set of responses that provide information—of much greater depth than questionnaire responses—from the group of people who are somewhat familiar with the topic. All points of view are accepted and disagreements are okay as long as they are over the ideas.

The focus group usually meets at a neutral site and is set up so that everyone is able to see everyone else. The sessions typically run 1 to 2 hours. A letter is sent to participants ahead of time with a description of the general purpose of the focus discussion. When the group meets, ground rules are set. General questions usually begin the focus group and the more detailed questions are saved for later. Everyone has an opportunity to contribute to each and every question. A recorder keeps track of all the comments. Often a flipchart is used to record the responses.

One purpose of a focus group is to gather data that is meaningful and that will be used in decisionmaking. Another use for the focus group is to measure progress, improvement, or a new approach to providing services.

As with any method of data collection, the focus group has its pluses and minuses. Some of the pluses are that a focus group can usually be done in a cost-effective manner. A lot of information can be generated, and solutions to existing problems, as well as needs for new programs, can surface. The information supplies you with impressions that cannot be gained any other way.

The NFPA Public Education Division, for example, while developing the new "all injury control" curriculum entitled "Risk Watch" conducted focus groups in ten different cities in the U.S. and Canada before even developing the actual lesson plans and material that would be incorporated into the curriculum for a pilot test. Their goal was to find out from classroom teachers what their thoughts were on a curriculum that was designed as an experiential-based learning program.

Questions were provided to the focus group leaders, and meetings were set up to ask the questions in a free-flowing manner, allowing teachers to build on each others' ideas. At each session there was a focus group leader and a recorder. The recorder's sole responsibility was to capture the thoughts of the teachers present at the meeting and to draft the responses for submittal.

By listening to the community and being open to reevaluating, retooling, and changing your paradigm, you will be better equipped to work with the permission givers and the people you serve. Conversely, if the community and decisionmakers have false information or understanding of what you do, it is essential that you be willing to take the time and effort to educate them.

You must sell your services, your products, and your issues to the public. Build a strong relationship with your community that will last for generations, not just until the next budget cycle or crisis. Don't take anything for granted.

> *"Don't go around saying the world owes you a living; the world owes you nothing; it was here first."*
> — Mark Twain.

Meeting in the Middle

After finding out what the customers want, you need to take an honest look at the differences between what you do or want to do measured against the perceived wants or needs of the community. One way to approach this exercise is to involve a task force of community members to assist you with the process.

This approach worked well for a school bond issue in a small community in Pennsylvania. The local school board identified the need for a new school. The projected cost for the project was $38 million for a first-class school that included all the latest technology. The citizens were duly concerned with such an expense. The school board formed a citizen committee to review the project and to make recommendations regarding what would be included in the project. When all the recommendations were completed, there were several proposals generated, with differing options and features for the school. The cost ranged from $20 million to the original $38 million. The community again reviewed the proposals and finally decided on, not the lowest cost project, but the one that best met its overall needs. Buy-in was achieved and the initiative was passed.

When the stakeholders are involved in decisions that affect them directly, there is a significantly increased chance of success. They become part of the solution instead of part of the problem. It was a win-win situation for everyone in that small Pennsylvania town. Whether your department is funded through taxes or contributions, it is important to involve as many people in the process as possible.

Part of the process to achieve middle ground is the ability to give and take. In reviewing fire department customer wants and needs against services provided and resources available, it is critical to involve the community. After all, it is its financial support that pays for the fire department services. Be honest and realistic and be willing to listen. Be open to change and exploring alternatives to providing the same, altered, or new services.

> *"It is not the strongest of the species that survive, nor the most intelligent, but the one most responsive to change."*
> — Charles Darwin.

What do They Need?

Do your customers always know what they need? The answer is: "No, not always." An example of this would be the survey that the NFPA completed in February 1996. This survey revealed that most Americans feel safest from fire where they are actually at greatest risk. Most Americans underestimate their risk of fire danger and too few take the simple, key fire safety steps that can save their lives.

A national telephone survey of 800 Americans was conducted. The majority of respondents felt "somewhat to very confident"' about fire safety in their own homes, yet home is where the risk of fire is statistically greatest. These same survey respondents also felt confident about safety from fire while in their cars, even though private cars are another place where the risk of fire is great.

These respondents felt less confident about fire safety in hotels. However, because of improvements in fire safety throughout the hotel industry, including greater use of automatic fire sprinklers and alarm systems, hotels are at a low risk of fire.

> *"Americans seem to have an unrealistic feeling of overconfidence in safety from fire and a lack of critical knowledge of the safest response to a fire."*
> — Meri-K Appy, Vice President, Public Education, National Fire Protection Association.[5]

This is where effective communication comes in to play. It becomes a key factor in providing exceptional, needed customer service, which includes letting the community know what you know they need to be safe.

Resources - What's Out There?

According to Webster's *Ninth New Collegiate Dictionary*, a resource is "a source of supply or support" or "a source of information or expertise."

Historically, for a number of reasons, many fire departments have not reached out to the resources in their communities nor have they become directly involved in their communities' activities. Reaching out to gain an understanding of the available resources and becoming involved in the activities of the community will help the fire service to better understand both the needs and wants of the citizens.

In Phoenix, the entire department is "encouraged to continue to actively participate in the community to provide for the safety and well being of the people who live in our neighborhoods." This is an ongoing commitment made by all and especially by the individual stations. They determine the needs of their particular area of town and plan ways to be involved as an integral part of their own special community and to learn about their community's specific needs.

Often your approach to your community resources needs to be reevaluated. For years the Phoenix Fire Department had a Public Information Officer who had become the standard for others to follow. Steve Jensen was a reporter turned fire spokesperson. He had respect for, and an intense relationship with, the media in Phoenix. Unfortunately, when Steve passed away, a big void was left to fill.

At the same time, the legacy Steve left was a view of the media as a partner in public education. To reestablish the partnership Steve had built, the fire department decided to hold an intensive training academy for members of the media. The academy required a commitment of a full 40-hour, 5-day week. The benefit for the media was the loan of protective gear and certification allowing the graduates to go inside the fire lines to photograph and report closer to the action. This type of training for the media had not been held before.

The media enthusiastically enrolled eight students. The training program included Fire Ground Operations, Fire Behavior, Fire Suppression, Search and Rescue, Hazardous Materials, Helicopter, Trench and Mountain Rescue, Emergency Vehicle Driving, Emergency Medical Services, the Flash Over/Back Draft Chamber, and Stress Management. The week-long academy ended with a graduation ceremony.

After the Media Academy, each journalist was certified and an explanation was given about rules governing their scene privileges. Each certified journalist received a bright green helmet, easily identifiable on the fireground or emergency medical scene.

These journalists now have a clearer understanding of what is done at a scene and why. One of the many positives coming from the week-long academy is that a much clearer public education message is being given to the community following an event. An example of this would be the reporter telling the many reasons why a working smoke detector is so necessary following a house fire where no working smoke detector had been found.

What a resource these journalists have become to the fire department and the community! Resources come in many forms; the journalists are only one example.

Fire departments need to realize that people from the community can be a resource. They can respond as the journalists do in Phoenix, or they can volunteer their services as they do for many departments across the country. Volunteers not only fight fires, they also help departments with their senior outreach programs, inspecting home day-care facilities, only to name a few examples. These support functions are so important to all fire departments.

Another resource could come in the form of a grant. There may be local, State, and Federal dollars that can be obtained by writing a grant. This money is available, yet many departments have not realized that there may be grants that could meet their specific needs. Probably the best first step in the process is to identify a grant writer or a person in your department whom you can send to a grant-writing training program. Many community colleges and universities offer grant-writing courses. Local libraries will have "how to" books on grant writing. Your city may have a grant writer on staff. Some school districts and many colleges now employ grant writers.

Some of what you need to know about your organization and its budget concerns in order to complete the grant process includes:

- Operating money needed,

- Equipment needed,

- In-kind support needed,

- Financial resources in your organization at this time,

- Whether you are seeking long- or short-term funding, and

- If the activity is long-term and the funding short-term, how will you continue to fund the activity?

Where Can You Find Grant Funds?

Federal grants—Look in the *Federal Register* or the *Commerce Business Daily* in your local library. Obtain USFA's *Federal Domestic Assistance* Catalog (FA-132) by writing: USFA Publications, 16825 South Seton Avenue, Emmitsburg, Maryland 21727, or via the Internet at http://www.usfa.fema.gov

State grants—Check with your State Health Department for Preventive Health Block Grant monies, your State Highway Safety office for alcohol, traffic safety and seat belt programs, and your State EMS office for EMS funds.

Local government grants—Check with your local government to determine what grant monies may be available.

Foundations and Corporations—Check with your local library for the *Foundation Directory, National Data Book of Foundations, National Directory of Corporate Giving,* and other similar publications.

Some local foundations may be too small to be listed in one of the larger directories. However, every foundation must file with the IRS an annual report on giving. That report can usually be found in local libraries.

Further information may be found in USFA's *Guide to Funding Alternatives for Fire and EMS* (FA-141). Write for a free copy, or visit USFA's Web site at the previuosly listed addresses.

Each of these resources will give you contact names, addresses, and phone numbers; information on types of projects funded in the past and funding levels; and the method for initial approach, such as a letter of inquiry or requesting a grant application package.

Equipment is another example of a resource. Many larger fire departments donate very good equipment to smaller departments, which make good use of the equipment.

Another resource is a person's talent. An example of this would be an artist who can design a public education brochure for a department at no or very low cost.

Community resources are plentiful and endless. Often it takes someone in the fire department to ask. Everyone involved benefits.

Before asking anyone or any organization for support, make sure that your approach is well planned. Develop the following information before you start requesting resources:

- Written program description,
- Who should work on this particular project,
- A timeline for any project,
- Job descriptions for both paid staff and volunteers,
- Cost estimates, and
- Anticipated answers to possible questions.

Preplanning always pays off. Whether you are going after money, grants, or volunteers, have a plan in place before seeking these resources.

Identify Strengths and Weaknesses

As we said at the beginning of this manual, the fire departments of today will need to conduct business in innovative ways in order to compete and survive in the future. Coloring outside the lines, business as unusual, paradigm shifts, nontraditional approaches, and inclusiveness are the buzzwords of the future. It is the fire department that can be honest about what it does and does not do well, and is willing to change and to stumble occasionally to meet the future needs of the community, that will succeed. However, strengths should not be allowed to become weaknesses. When organizations are doing the wrong things—even if they do them flawlessly—the organization will have trouble coping and competing. Strengths become weaknesses when the environment changes, but behavior doesn't. Don't let the organization be locked into a set of skills that are or may become outdated.

Fire departments will need to become quick-change artists and be able to manage perpetual motion. One of the weaknesses of the fire service is its resistance to change. Just when you think you understand the situation, what you don't understand is that the situation just changed. Many fire prevention officers are locked into doing things just as traditionally as those who conduct suppression.

In evaluating what you do, you must be prepared to look critically at your organization at every level. You must be willing to let go, to change or alter the things that you do. In some cases, you may want to have a funeral for the old ways as you begin the new.

An example of this comes from a large State Fire Marshal's Office. The office determined that the State inspection reporting and data collection system had to be changed and updated. It was determined that the best way to proceed would be to start from scratch. Teams of inspectors were invited to assist in identifying critical information elements, output reports needed, and a process for determining compliance after an inspection.

The new system was developed and pilot tested to see if it would work before implementation. By the time the new system was ready to be fully implemented, rumors of the complexity of the new system had erupted division-wide, signifying uneasiness with the change.

The day of the implementation training, the facilitator handed each inspector a set of five old forms, with all the redundancies that had to be collected in the old system. They were also taped to the walls on both sides of the rooms. A funeral was then held for the old system and its cumbersome requirements. At the end of the activity, an imaginary switch was turned off, facilitators asked for a moment of silence, and inspectors were instructed to tear the old forms into as many pieces as possible and discard them into the wastebasket.

The facilitators handed out the two new forms (down from five), and the new system was explained. They also explained that for the next six months the system would be monitored and each inspector would be able to suggest changes. At the end of the presentation, the 23 members of the inspection team were in agreement with the new concept and system. After six months, a few minor changes were suggested and implemented, and the program is working well today. The new system provides a tremendous amount of new information and provides the inspection unit with more time in the field as opposed to all the time spent previously on paperwork.

Finding the strengths and weaknesses in what you do will generate change, and change will generate uneasiness. How you address the change will mean the difference between success and failure. Whether in suppression activities, enforcement, or public education activities, fire service leaders must be willing to risk examining what is and is not working and determine if it's the right thing to be doing, even if it is done very well.

Planning: Bury the Dinosaur

Fire departments, even as government agencies, must operate as a business and do what they do well—and do it better than their competition.

Building a reputation for quality service can't happen if your internal customers — your employee—do not believe in your philosophy of customer service. A strategy must be devised to win over your internal customers. Your internal customers must become part of the team. Quality is not just the job of the managers; it involves everyone in the organization.

The Ritz-Carlton, 1992 winner of the Malcolm Baldrige National Quality Award, credits its receiving the award to its employees. Coaching the employees in the Ritz-Carlton way of life starts even before they begin working there. Every applicant is interviewed three times before receiving a job offer. After being hired, the new employee spends two days in an orientation class. Personal interaction and company philosophy are stressed. Ritz-Carlton employees receive 100 hours of quality training each year.

Successful internal marketing has as its goal the building of successful service strategies which can serve the external customer. Being able to deliver consistent, high-quality service requires the effort of everyone in the organization. This is why listening to your internal customers is as important as listening to your external customers. Surveys and focus groups with the internal customer are as helpful as they are with the external customer.

A wonderful example of listening to both your internal and external customers is found in the 1996 Reedy Creek Emergency Services Strategic Plan. Reedy Creek Emergency Services Department is located in Lake Buena Vista, Florida. Representatives of upper and middle management along with labor representatives were brought together to participate in the development of a strategic plan. A cross section of customers of Reedy Creek Improvement District, representing The Walt Disney World Company, rated the services and identified the strengths and weaknesses of the organization from their perspectives. Both the customer concerns and the needs of the organization were addressed.

The process selected for the development of the strategic plan is described as follows:

> "The process allowed for community input as well as department participation. It focuses on community needs and was driven by the central theme of 'What's in the best interest of the customers served.' Customer-centered strategic planning, unlike traditional planning processes, is designed to focus on programs and identify issues from the customer up rather than the mission statement down."

Through this process, six goals were developed. Briefly, they are:

- Develop and implement a long-range plan for the needs within the Reedy Creek Improvement District Emergency Services;

- Develop and implement a program for improving interdepartmental teamwork and relationships;

- Develop and implement appropriate administrative and operational policies and guidelines;

- Design and implement Reedy Creek Emergency Services pre-incident communication and emergency plans to aid in reducing the number of incidents that occur within the community;

- Enhance Reedy Creek Emergency Service's customer relations; and

- Design, implement, and/or revise a career development, education, and training program to ensure that all employees have the opportunity to be as well prepared as possible and also have the ability to succeed.

The process allowed for both internal and external input. It focused on the needs of the community and was driven by the theme of "What's in the best interest of the customers?"

Fire and emergency services are now in a very competitive cycle. Customer demands are increasing while dollars and adequate staff are shrinking. The process that the Reedy Creek Emergency Services Department went through has enabled it to plot a course for planned change and improvement as well as the necessary tools needed to accomplish the tasks.

See Appendix C, "A Step Back Helps Us Move Forward," for a description of how the Paso Robles, California Fire Department changed its focus internally.

Competing Through Promotion, "Promoting to Compete"

Today and in the future, fire departments are and will be competing for scarce resources, while at the same time taking on more responsibilities. How do we compete? We compete by promoting what we do, the services we provide, educating citizens about the fire problem, and, most importantly, by informing the public of the value of our service. All of this should be based on the community's needs and wants.

This is a public relations and motivation activity. If the community has not been involved with the fire department in the past, now is the time to change how you interact with them. This change should not be feared, it should be welcomed. The goal, after all, is survival. Fire departments may be unable to continue operating as a lone wolf and survive in the future. This is an organizational change of significant importance.

> *"Combine your efforts seamlessly with others who, though very different from you, are contributing to the same end results."*
> — *Anonymous*

In competing we need to take a lesson from advertisers to sell our services and products and begin to mirror some of their approaches. Let's look, for instance, at how Nike advertises athletic shoes. The company invests significant resources in researching comfort and safety issues for the wearer. This is valuable for product development but it's not the way they sell shoes. Instead, they motivate the public to buy their product by promoting the adventures and the places the shoes could take them and by tying their product to use by sports personalities.

What can we learn from advertisers? Simply this: We need to view what we do differently and learn from others. We are so used to selling the public on the idea of death and destruction, consider what would happen if we sold them on the concept of quality of life, of the future in terms of generations, and of living to be 100 years old.

There is no lack of creativity in the fire service. However, we must be willing to think creatively, to look for the possibilities instead of obstacles. If you were shown a picture of an acorn, for example, what would you say could be done with it if you had but one to use? If you said you would feed a squirrel, you flunk. Think harder. If you can imagine using it to grow a tree, you get a C. If you said it has possibilities to grow a forest then you get an A. By eliminating the traditional limiters of the fire service paradigm, you can be open to possibilities beyond expectations, not only for the operations of your fire department, but for your public relations and public education efforts. There are several excellent books on creativity and innovation listed in the reference section of this manual.

Don't be afraid to try something that has never been tried before. Think, create, and plan with the future in mind. More people, more technology and tools, and more knowledge add up to a future where the shelf life of solutions will become history in a hurry. It is essential to give the culture of the fire service permission to change today, so that the organization can survive in the world of rapidly advancing history.

> *"Computer power is now 8,000 times less expensive than it was 30 years ago. If we had similar progress in automotive technology, today you could buy a Lexus for about $2. It would travel at the speed of sound and go about 600 miles on a thimble of gas."*
> — *John Naisbitt, Author of Global Paradox.*[18]

Keep reaching and stretching to undo yesterday. Continue to improve bit by bit, and eventually small gains will add up. What we consider "good" today may be seen as "so-so" tomorrow.

Promoting and marketing your department will take the talents of many people. You need to look to and convince your community to help you. If you are able to promote your services in a way that solicits a positive response, you have been successful with your efforts. The more colorful and unique you are with your approach, the less competition you will encounter and the more memorable your message will be. When the time comes to divvy up scarce resources, you will have a prominent place in line.

Egos cannot play a role in promoting your department and services. Be visible, be a friendly, willing participant in community activities, and become known as a helper and a resource for all occasions. Fire departments that view this as too much work may end up on the sidelines, watching the competition do what they should be doing. Attitude, as discussed in the beginning of this manual, means everything.

Extinction shouldn't be an option for any fire department. If you have a problem today, you must be ready to react immediately. In the past, competition wasn't as stiff and there was more time for recovery. Those days are gone forever. Problems can't be ignored, and there are no permanent solutions. We need to develop faster reflexes and provide an entirely new set of responses or solutions for every problem. If something doesn't work out, don't wait for the problem to disappear, try it another way. Don't give up. Winners know the secret — get right back up and try again. The bottom line is to give our investors (the community) a good return on their investment.

> *"There is no off-season anymore and when you win, everything feels good!"*
> — Anonymous

Community-Based Promotion

The fire department needs to know both what the community wants and needs before it can determine the messages it wants to get out to the community.

In 1978, the Phoenix Fire Department became a test site for the National Fire Protection Association's (NFPA) Learn Not to Burn™ Program. Over the years, with the help of school teachers across the city, the department provided valuable fire safety information through this program to thousands of Phoenix schoolchildren.

At the same time, the department's firefighters became concerned that many preventable, nonfire injuries and deaths were occurring. A drowning, a motor vehicle accident injuring occupants who were not wearing seatbelts, a child who drinks something harmful, or someone hurt in a crosswalk were some of the calls to which firefighters were responding.

As a result of their concerns, firefighters met for almost two years to develop a new curriculum to address the types of calls to which they found themselves responding. With the help of injury prevention experts from the community, they developed a new curriculum called the Urban Survival Program.

The overall goal of this new Urban Survival Program is to teach schoolchildren, adults, elderly, and the disabled the skills necessary to protect themselves and their families by responding promptly and effectively when confronted with a fire or life safety hazard.

Because the department chose to listen to its internal customers, a curriculum is being taught in the community which is having a direct impact on drowning, pedestrian safety, accidental poisoning, babysitters, desert survival, firearm safety, CPR awareness, pet safety, outdoor recreation safety, latchkey children, and injuries at construction sites.

The Urban Survival Program is a joint partnership between the fire service (the Phoenix Fire Department and the United Phoenix Fire Fighters Local 493) and the educational community. It involves teachers, firefighters, PTA members, and organizations such as the Phoenix Children's Hospital, the Maricopa Medical Center Burn Center, Shriners, and Good Samaritan Poison Center. With all these community partners, the program has become a winner.

Youth firesetting is not just the fire department's problem, it is a community issue. Based on this thinking, the Phoenix Fire Department put together a Community Advisory Panel to combat the youth firesetting problem. With three fire deaths in 1995 and two in 1996 as a direct result of fires set by youth, and $1 million of structural damage due to such fires each year, this problem is most definitely a major one.

The goals of the Community Advisory Panel are to use the members' collective resources and expertise to reduce the youth firesetting problem; to share with one another various programs to see if any could help reduce youth firesetting; and to increase the Community Advisory Panel members' agencies awareness of the Youth Firesetting Prevention Program.

This group of community leaders meets quarterly. They represent the Attorney General's Office, Child Protective Services, a children's magazine, an insurance agency, the juvenile justice system, Maricopa County Attorney, Maricopa County Public Defenders, Paradise Valley School District, St. Luke's Behavioral Health Center, Maricopa County Medical Center, and the United Phoenix Firefighters Local 493.

> *"The United Phoenix Area Firefighters Association has always worked closely with fire department management in developing and running programs that are innovative and responsive to the needs of the citizens we serve. We view these public service activities as an integral part of our union's role in the community."*
> — Pat Cantelme, President of the United Phoenix Area Firefighters, IAFF Local 493[18]

Many doors have been opened to the Phoenix Fire Department because of the members' knowledge, contacts, and commitment to help with this problem of young people setting fires. The department has put together a model program that is able to use resources that weren't even known about prior to the Community Advisory Panel's first meeting.

Reaching out to the community is smart.

A department's major objective must be to provide its customers with the highest possible level of customer service. This entails knowing the needs and wants of both its internal and external customers. This type of quality customer service has to be the responsibility of everyone in the organization in order for it to work continuously and successfully.

It takes leadership to successfully guide a department through the inevitable and ongoing change brought about by factors like demographic shifts, increased scarcity of resources, and a public with higher expectations. The lessons of this manual tell us first, to recognize and accept that change will happen. Second, that there must be a vision for how a fire department will thrive with the change. Third, that this vision must affect an internal paradigm shift. And, fourth, that this paradigm shift must be driven by the real and perceived needs of the customer. All of this will happen in an increasingly political arena where local appointed and elected officials—important fire department stakeholders—will make difficult decisions about how to allocate scarce resources among all local government departments. These departments all serve the same customer and all face the same changing environment as the fire department.

What are some of the steps that may be taken to thrive in an environment of change?

1. Review your mission statement. Is it broad enough? Does it acknowledge the necessity for prevention activities? For PIER activities?

2. Assess how your department is doing with services currently being provided. What kind of standards does your department hold itself to? How is this evaluated?

3. Do your homework. Look at available data from sources such as the U.S. Census Bureau, USFA NFIRS program, State and local agencies, and others listed in Chapter 3; look at neighboring and/or comparable departments.

    Talk to your customer. Conduct focus groups. Distribute surveys. Gather information about the perceived needs and satisfaction levels of your clients. Chapter 4 discusses methodologies you can implement to accomplish this.

    If you feel that you don't have the expertise to analyze the data, go to outside sources. Many universities have students who get credit for completing outside data analysis or program evaluation projects.

    Data available from a variety of sources and information directly from your customers will give you what you need to make sound decisions about your services — what is no longer needed, what may be improved, what is lacking — and to make your case to the local elected and appointed officials who ultimately allocate the resources.

Steps 1 to 3 should help you to determine whether your mission statement and the services currently provided are appropriate, given the real and perceived needs of your community and how well your department is meeting those needs. These steps should help you identify weaknesses within the department and any unmet community needs. Perhaps your department's response performance is outstanding, but citizens are unaware of this unless they have been a victim of fire. This indicates the need for outreach to the community. The Salida, Colorado example in Chapter 1 shows an effective approach to outreach activity—work with the media, make presentations to civic groups, etc. By the same token, the department whose response performance is outstanding may have a real problem with high-risk behaviors in the community—a lack of smoke detectors, children playing with butane lighters. Perhaps it's time to redouble your public education efforts and change behaviors.

Fill the service gaps that you've identified. Can you add or improve services with existing resources? If additional resources are required, how will you secure them? Look for partners with a common objective. Children's Village in Washington County, Maryland (Chapter 4), is an excellent example of a coalition of community organizations cooperating to achieve a common goal. There are many other fine examples of coalitions that are less elaborate, but also effectively meet a fire department objective shared by another organization. Child safety, for example, is a goal shared by many organizations. Also in Chapter 4 are suggestions for identifying grant funding sources. Be creative!

Finally, leverage the fine services that you do provide. Each response to an emergency, each visit to a classroom, each community CPR class is an opportunity to showcase the department. Take every possible opportunity to make your customer aware of what you do.

WEB SITE DIRECTORY

Locating Resources via the Internet

The information available on the World Wide Web is growing at an astounding rate. Even the most casual "browser" can easily find basic information about a subject. An experienced user can extract reams of information and resources, sometimes very specialized, about a subject.

Fire service leaders seeking information and resources to support local efforts such as PIER programs can start their search on the Web. A wealth of useful Web sites will help anyone find sample programs, information resources, publications, professional development opportunities, funding sources, potential local partners of national groups, and much, much more.

Experienced Internet users know that a search for applicable Web sites can be performed through one of the many "search engines" available on the Web. To aid both inexperienced and experienced users in getting started, however, the following list of useful Web sites has been assembled. These sites offer a wealth of information and resources, and are the logical starting point for a fire service leader searching for additional sites, organizations, and resources in the areas of:

Public information
Public relations
Public education
Media relations
Leadership
Funding
Partnership/Coalition building

Brief information about some of the resources available through the organizations that sponsor these Web sites is described. Readers are encouraged to browse each site to locate and take advantage of the full range of information and resources that a site or an organization may provide.

The following listing of Web sites represents a few examples of resources available. It is not intended to be all inclusive. Use these sites as jumping-off points for further research on the Web. A treasure trove of information can be opened with just a little additional searching.

American Red Cross (ARC)

www.redcross.org

To locate the address and the telephone number of the local Red Cross chapter serving your area or nearest to you, use the search-by Zip Code feature under Local Chapters on the American Red Cross Web site.

The American Red Cross provides relief to victims of disasters and helps people prevent, prepare for, and respond to emergencies. The ARC offers a wealth of materials and publications on how people can prepare for and respond to fires and other disasters.

Sections on the ARC Web site of particular interest to browsers from the fire service are the Disaster Services, Health and Safety Services, and Youth Involvement sections:

- **Disaster Services**—Includes sections on After a Disaster Strikes and Disaster Safety Information. For example, the Disaster Safety Information section offers individuals information on preparing their home and family for any kind of disaster. For each type of disaster (fire, earthquake, flood, for example), materials available through the Red Cross and its chapters are listed and described. Materials include brochures, educational kits, posters, coloring books, and videos.

- **Youth Involvement**—This section offers numerous avenues for youth involvement in community-oriented disaster and other services.

Fire-EMS Information Network

www.fire-ems.net/

This commercially supported site is notable for its extensive number of links to local fire departments and firefighters' locals. With a little work, an interested browser can locate numerous examples of programs being conducted by local fire departments and which may be transferrable to your department.

Fire Protection Publications (IFSTA)
International Fire Service Training Association

www.fireprograms.okstate.edu/index.ssi

Oklahoma State University
930 North Wills
Stillwater, OK 74078-8045
(405) 744-5723
(800) 654-4055
FAX: (405) 744-8204

IFSTA is a nonprofit educational association dedicated to firefighting techniques and safety through training. Fire Protection Publications (FPP) publishes training texts, and researches, acquires, and markets learning and teaching aids. The FPP's online catalog offers information on all of its products, including:

- **Public Fire Education Digest**—A subscription newsletter that is a compendium of public fire education news from the U.S. and Canada.

- **Fire and Life Safety Educator's Resource Kit**—Materials and resources that may be used by a fire and life safety educator.

New on the Web site is a section for Fire and Life Safety Resources. It provides links to products and information on public education.

The Foundation Center

www.fdncenter.org
79 Fifth Avenue
New York, NY 10003-3076
(212) 620-4230
FAX: (212) 691-1828

Fire service managers looking for nongovernmental foundation funding should start here —The Foundation Center Online. The center is a nonprofit information clearinghouse that provides information on foundations, corporate giving, and related subjects.

Start at the Foundation Center Web site to learn more about funding sources, training and information services (such as how to write a proposal), links to funding sources on the Internet, and available directories of foundations and other publications.

International Association of Fire Chiefs (IAFC)

www.iafc.org
4025 Fair Ridge Drive, Suite 300
Fairfax, VA 22033-2868

The IAFC provides leadership to career and volunteer chiefs, chief fire officers, and managers of emergency service organizations. Of primary interest on the IAFC web site is the subscription-only service known as the ICHIEFS Private Network. It is a messaging and discussion group system that allows private conversations by fire service leaders on a wide range of topics. The private network can also be accessed at **www.ICHIEFS.org**

The IAFC's Management Information Center offers publications that are compiled information packets. Of interest to those using this manual are *Recreating the Fire Service*, a publication designed to help improve the quality of a fire department's service through a team-building approach, and *Volunteer Fire Chief's Sourcebook*, a two-volume, self-help guide for the volunteer service leader.

International Association of Fire Fighters (IAFF)

www.iaff.org
1750 New York Ave., NW
Washington, DC 20006
(202) 737-8484
FAX: (202) 783-4570

The IAFF maintains an extensive Web site that offers a wealth of information about the labor organization for firefighters and EMS personnel. Fire service professionals will find numerous useful tools on the site, including extensive updates on legislative and regulatory issues, health and safety information, EMS tools, news bulletins, official statements, upcoming events, and links to other sites.

International City/County Management Association (ICMA)

www.icma.org
777 North Capitol St., N.E., Suite 500
Washington, DC 20002-4201
(202) 962-3600
FAX: (202) 962-3500

ICMA publishes and distributes hundreds of publications and other resources that are valuable for anyone in local government. The Association's Web site has an online publications catalog that provides a list of available products. Among the areas of local government management covered by ICMA's publications are Community Relations, Finance, Management, Police and Fire Services/Public Safety, and Service Delivery Management.

ICMA offers professional development through self-study courses that can be taken by groups or individually. Courses include The Citizen as Customer, Managing Fire Services, and Emergency Management Principles and Practice for Local Government.

Government Services Television Network, sponsored by ICMA and other local government organizations, provides video-based training programs on a wide range of topics.

The ICMA Web site also is a good starting point to locate other local government Web sites and organizations.

National Fire Protection Association (NFPA)

www.nfpa.org
1 Batterymarch Park
Quincy, MA 02269-9101
(617) 770-3000
FAX: (617) 770-0700

The mission of the NFPA is to reduce the burden of fire on the quality of life by advocating scientifically-based consensus codes and standards, research, and education for fire and related safety issues. The NFPA web site has the latest information on NFPA activities, publications, seminars, and educational programs. The site includes sections on NFPA seminars and meetings; the NFPA codes and standards-making process and technical committee information; codes and standards products; and periodicals.

Sections that can provide visitors with a variety of information and resources are:

- **Fire Safety Information**—Contains useful information on fire safety, such as Home Fire Safety Tips, Seasonal Fire Safety Tips, and Fire Protective Clothing.

- **Media Access**—Contains sample public service announcements, news releases, current fire safety information, information on Fire Prevention Week, and access to NFPA spokespeople.

The NFPA site also has an extensive collection of links to other fire-related Web sites. Look under "Fire Resource Links."

National Safe Kids Campaign

www.safekids.org
1301 Pennsylvania Ave., N.W., Suite 1000
Washington, D.C. 20004-1707
(202) 662-0600
FAX: (202) 393-2072

The National Safe Kids Campaign Online site will lead interested browsers to a wealth of resources on safety—including fire safety—for kids. The site is useful for locating safety-related organizations and associations not usually found on fire-related sites.

Use these sections to locate resources that will be useful in local fire service programs:

- **Resource Catalog**—Lists materials available for purchase; includes injury prevention, bike helmet and bike safety, fire prevention and fire safety, and child occupant protection.

- **State and Local Coalitions**—By State, lists cities and counties with local Safe Kids Coalitions. Some listed have e-mail links. To locate the address and phone numbers of most, call the national headquarters.

- **Health/Safety Related Links**—Useful safety and injury-prevention links not found on fire-related Web sites.

National Safety Council (NSC)

www.nsc.org
1121 Spring Lake Drive
Itasca, IL 60143-3201
(630) 285-1121
FAX: (630) 285-1315

The NSC provides training, education programs and materials, and consulting to help educate and influence society to adopt safety, health, and environmental policies, practices, and procedures that prevent and mitigate human suffering and economic losses arising from preventable causes.

The NSC has a wide array of programs, council divisions, products and services, publications, and training that PIER programs can tap into. To access information on the NSC's many resources, click on "Web Site Directory" on the NSC Web site.

Connect with local council chapters by locating the one serving your area under "Council Chapters." Addresses and phone numbers of all chapters are listed; many also have Internet connections listed.

National Volunteer Fire Council (NVFC)

www.nvfc.org
1050 17th Street, N.W.
Washington, D.C. 20036
(202) 887-5700
FAX: (202) 887-5291

The NVFC is a nonprofit membership association representing the interests of volunteer fire, EMS, and rescue services. The NVFC Web site addresses the needs and concerns of volunteers and volunteer departments.

Among the information found on its Web Site, the NVFC offers a Resources and Services section, which provides compiled listings of resource materials such as manuals, books, pamphlets, videos, and reports. The listings include sections on Prevention and Public Education, Funding, Training, Volunteer Issues, and Retention and Recruitment.

The NVFC also offers a manual on information and ideas to start programs in public fire education. *Public Fire Education in the Volunteer Fire Service* offers examples from various departments. Contact the NVFC to order.

State and Local Government on the Net

www.piperinfo.com/state/states.html

This Web site is a "virtual" resource that is a gateway to resources within your State. This site connects you to the Web sites of the legislative, judicial, and executive branches of State government, along with boards and commissions and many regional commissions. It also features links to counties and cities that operate Web sites, as well as State associations of counties and cities.

United States Fire Administration (USFA)

www.usfa.fema.gov
16825 South Seton Avenue
Emmitsburg, MD 21727

The place to begin any online search for resources to support a local program is USFA's Web site. On the site, interested viewers should click on these sections:

- **Publications**—The USFA online publications catalog contains more than 300 re sources. Nearly 60 of these are public education program materials. All items may be ordered online or through the USFA's Automated Publications Ordering line at (301) 447-1660. Some items can be downloaded from the Electronic Publications Refer ence Page.

- **National Fire Academy**—The material here covers the wide range of courses and programs offered by the NFA. Check out opportunities for on-campus courses offered at Emmitsburg, as well as distance-learning options, such as the Regional Delivery Program.

- **National Fire Programs**—This section provides details and resources on USFA's many programs, including its Fire Safety and Public Education efforts. Provides information on USFA's Public Education Partnerships (and how to obtain further resources from each partner); the Quick Response Unit (which consists of a series of fact sheets and print and broadcast public servie announcements).

- **<u>Learning Resource Center</u>**—Users of the Web site may access the LRC's Online Card Catalog to perform a literature search. The LRC has a collection of more than 50,000 books, reports, periodicals, and audiovisual materials. Details on borrowing materials are included on the Web site.

REFERENCES

Scott Baltic, "Hands Across the Water," *Fire Chief*, Vol. 40, No. 10, pp. 36-47, October 1996.

Scott Baltic, "All the Learned and Authentic Fellows," *Fire Engineering*, Vol. 40, No. 12, pp. 24-29, December 1996.

Warren Bennis, On Becoming a Leader, 1989, Addison-Wesley, New York.

Ken Blanchard, "Shower People with Information," *Executive Excellence*, Vol. 12, No. 4, pp. 11-12, April 1995.

Alan Brunacini, "The Gospel According to Phoenix," *Fire Chief*, Vol. 40, No. 8, pp. 52-56, August 1996.

John W. Cebrowski, "Ten traits of High Performance," *Executive Excellence*, Vol. 12, No. 5, pp. 18-19, May 1995.

Community Education Leadership, 1996, National Fire Academy, Emmitsburg, MD.

Daryl R. Conner, Managing at the Speed of Change, 1995, Villard Books, New York.

Patrick E. Connor and Linda K. Lake, Managing Organizational Change, 2nd edition, 1994, Praeger, Westport.

Stephen R. Covey, "Faith in the Future," *Executive Excellence*, Vol. 13, No. 5, pp. 14, May 1996.

Robert A. DiPoli, "End the 'Burning Baby' approach," *Fire Chief*, Vol. 39, No. 8, pp. 68-70, August 1995.

Peter Drucker, "Innovation Imperative," *Executive Excellence*, Vol. 13, No. 12, pp. 7, December 1996.

Nancy K. Grant, Ph.D. and David H. Hoover, Ph.D., Fire Service Administration, 1994, National Fire Protection Association, Quincy, MA.

David Gratz, Fire Department Management: Scope and Method, 1972, Glencoe Press, Beverly Hills, CA.

Alan Greenspan, "Forces Driving Our Economy," *Executive Excellence*, Vol. 13, No. 12, pp. 16-17, December 1996.

Chuck Haga, *Minneapolis Star Tribune/Twin Cities Journal*, December 29, 1996.

Ronald A. Heifetz and Donald L. Laurie, "The Work of Leadership," *Harvard Business Review*, pp. 124-134, January-February 1997.

Howard E. Hyden, "From Manager to Leader," *Executive Excellence*, Vol. 11, No. 12, pp. 10, December 1994.

John P. Kotter, "Leading Change: Why Transformation Efforts Fail," *Harvard Business Review*, pp. 59-67, March-April 1995.

Ken Lavoie, "Justifying your Department's Existence is a Full-Time Job," *Fire Chief*, Vol. 39, No. 8, pp. 60-67, August 1995.

Leadership in Public Fire Education: The Year 2000 and Beyond, United States Fire Administration, Emmitsburg, MD.

Thomas Peters and Robert Waterman, Jr., In Search of Excellence, 1982, Harper & Row, New York.

Thomas Peters, The Pursuit of WOW!, 1994, Vintage Books, New York.

Price Pritchett, The Employee Handbook of New Work Habits for a Radically Changing World: 13 Ground Rules for Job Success in the Information Age, 1996, Price Pritchett Publishing Company.

Price Pritchett and Don Pound, A Survival Guide to the STress of Organizational Change, 1995, Price Pritchett Publishing Company.

Reedy Creek Emergency Services Department 1996 Strategic Plan.

Brian G. Robinson, "Pressures for Change," *Fire Chief*, Vol. 4, No. 2, pp. 111-113, February 1996.

Dennis Rubin, Wingspread IV Report (draft), 1996, Dothan, AL Fire Department, Dothan, AL.

Peter Senge, The Fifth Discipline, 1990, Currency Doubleday, New York.

Shaping the Future, 1996, National Fire Academy, Emmitsburg, MD.

Strategic Management of Change, 1996, National Fire Academy, Emmitsburg, MD.

Patrick L. Townsend and Joan Gebhardt, "Leadership at Every Level," *Executive Excellence*, Vol. 11, No. 12, PP. 13, December 1994.

[1]Bill Peterson, Chief, Plano, Texas, Fire Department, Interview, December 22, 1996.

[2]Peter Drucker, "Innovation Imperative," *Executive Excellence,* Vol. 13, No. 12, pp.7, December 1996.

[3]Ed Kirtley, Chair, NFPA 1035 Committee for Fire and Life Safety Educator, Interview, May 22, 1997.

[4]Leadership in Public Fire Safety Education: The Year 2000 and Beyond, United States Fire Administration, Emmitsburg, MD.

[5]Meri-K Appy, Vice President, Public Education, National Fire Protection Association, Interview, December 11, 1996.

[6]Dennis Rubin, Chief, Dothan, Alabama, Fire Department, Interview, December 22, 1996.

[7]Alan Greenspan, "Forces Driving our Economy," *Executive Excellence,* Vol. 13, No. 12, pp. 16-17, December 1996.

[8]Alan Brunacini, "The Gospel According to Phoenix," *Fire Chief*, Vol. 40, No. 8, pp. 52-56, August 1996.

[9]Frank Carter, Deputy Chief, Colorado Springs, Colorado, Fire Department, Interview, December 22, 1996.

[10]John P. Kotter, "Leading Change: Why Transformation Efforts Fail," *Harvard Business Review,* pp. 59-67, March-April 1995.

[11]Howard E. Hyden, "From Manager to Leader," *Executive Excellence*, Vol. 11, No. 12, pp. 10, December 1994.

[12]Ken Blanchard, "Shower People with Information," *Executive Excellence*, Vol. 12, No. 4, pp. 11-12, April 1995.

[13]Peter Senge, The Fifth Discipline, 1990, Currency Doubleday, New York.

[14]Ronald A. Heifetz and Donald L. Laurie, "The Work of Leadership," *Harvard Business Review*, pp. 124-134, January-February 1997.

[15]Stephen R. Covey, "Faith in the Future," *Executive Excellence*, Vol. 13, No. 5, pp. 14, May 1996.

[16]Price Pritchett, The Employee Handbook of New Work Habits for a Radically Changing World: 13 Ground Rules for Job Success in the Information Age, Price Pritchett Publishing Company, 1996.

[17]Tom Hanify, President, Indianapolis Metropolitan Professional Firefighters Local 416, 1998 Phoenix Fire Department Conference.

[18]John Naisbitt, Global Paradox, 1994, The Hearst Corporation, New York, NY.

[19]Pat Cantelme, President, United Phoenix Area Firefighters, IAFF Local 493, 1998 Phoenix Fire Department Conference on Change

APPENDIX A
WINGSPREAD IV
Statements of Critical Issues to
The Fire and Emergency Services in the United States

Wingspread: The Name

Like many successful conferences that have had lasting importance to the nation's fire service, such as the Williamsburg '70 Conference, the Stonebridge Conferences, the Rockville Report, etc., the name often associated with such meetings and their post conference reports refer to the location where a given conference was held, be it a city or the name of the conference center itself. Such was the case with the original Wingspread Conference.

Named for the Wingspread Conference Center, owned by the Johnson Foundation in Racine, Wisconsin, the original conference was held there in February 1966. Designed in 1938 by the famous architect Frank Lloyd Wright, about the Wingspread House, also called the Johnson House, Wright would later be quoted as saying, "Spread its wings, it did." The Johnson's were of the Johnson Wax family. The house was made into a conference center in 1960, and has since been host to thousands of nationally and internationally renowned meetings on important issues to society. The Wingspread Conference Reports on fire in America are among their greatest ventures which they hold dear to this day.

The Johnson Foundation takes great pride in the continuing use of the Wingspread name by those meeting in Dothan, Alabama, in October 1996, for the Wingspread IV Conference. Interesting, as well, is the name, Wingspread, and the logo, "Hotfoot," of the Federal Fire Programs, where the Federal Eagle is shown stamping out fire, are intertwined connection of importance. From the Wingspread Reports much was drawn toward the establishment and evolution of today's U.S. Fire Administration and National Fire Academy.

Those Who Have Gone On Before

Wingspread is unparalleled in fire service history. Those who have made up each of the first three Ad Hoc Committees were truly giants in the fire service. And, certainly, no one name stands out more than William E. Clark, who was the inspiration behind each of the conferences and who was actually the person who originally got the effort established. Bill, who died this past year, had served with distinction in the New York City Fire Service Department, had also been an industrial fire chief, a state fire service director of the fire service training, a county fire chief, an organizer of a major national fire service instructors group, and a respected writer and lecturer. While serving in Wisconsin, he approached the Johnson Foundation about sponsoring a gathering of national fire service leaders to study and report on the nation's fire problems and the steps needed to improve the fire services in America.

Others who participated in the Wingspread Conference of 1966, 1976, and 1986, have been described as being among "America's Who's Who in the Fire Service." Each of the participants have been people who cared deeply about the future of the fire service and had a passion for improving both educational opportunities for fire personnel and the fire loss picture in this country.

1996: The Recent Opportunity

As the fourth Ad Hoc Committee of dedicated individuals gathered in Dothan, Alabama, for Wingspread IV, their challenge was no less great and their final product no less important than any previous effort. The next decade will surely be as complicated and difficult as any in history. This group has before it, the excitement and optimism of all who went before, and opportunities to make a difference in the lives of those yet unborn. The spirit of Wingspread continues to motivate and reinforce the best in all of us, and builds upon the dreams and hopes of a better tomorrow.

Introduction

The American fire service is the response resource of first and last resort when communities and individuals are confronted by sudden and unexpected calamities of modern life. This includes a public expectation for quick and efficient emergency medical services. Leaders of the fire and emergency services should not allow these essential roles and responsibilities to be trivialized by the argument that the frequency of reported fires has been reduced.

The population of the United States has nearly doubled since the Wingspread processes began in 1966, accompanied by proportional increases in hazardous materials manufacturing, refining and chemical processes, and the transportation of people and hazardous materials in all manner of conveyances. The transportation infrastructure is increasingly fragile and subject to collapse. There is a greater density of population in most urban areas, at a time when acts of terrorism and natural disasters are becoming more commonplace. Fire departments must provide comprehensive training and education which include fire prevention, disaster preparedness, emergency medical care, safety, and hazardous materials awareness, specific to the customers and their community.

There was a deliberate effort to add divergent interests and views to the Wingspread IV Conference. This document reflects those diversities; the reader will recognize the differences. Certainly there is no single solution for many of the issues identified in this report. In their discussion, the participants did not urge uniformity but rather acceptance of the many diversities in the business. Throughout the meeting, various issues and considerations were occasionally initially characterized as being exclusively career or volunteer in nature. During the ongoing discussions it became apparent that the concerns, issues, and challenges were really true for all emergency service organizations.

However, it must be acknowledged that there are differences in the personnel management and deployment of career, combination, and volunteer fire services. Each fire department must define its capabilities and educate its customers of reasonable expectations. The highest levels of service should be the challenge of every fire service organization.

Emerging Issues of National Importance

1. **Customer Service:** The fire service must broaden its focus from the traditional emphasis on suppression to a focus on discovering and meeting the needs of its customers.

2. **Managed Care:** Managed are may have the potential to reduce or control health care costs. It also will have a profound impact on the delivery and quality of emergency medical services.

3. **Competition and Marketing:** In order to survive, the fire service must market itself and the services it provides, demonstrating to its customers the necessity and value of what it does.

4. **Service Delivery:** The fire service must have a universally applicable standard which defines the functional organization, resources in terms of service objectives (types and level of services), operation, deployment, and evaluation of public fire protection and emergency medical services.

5. **Wellness:** The fire service must develop holistic wellness programs to ensure that firefighters are physically, mentally, and emotionally healthy and that they receive the support they need to remain healthy.

6. **Political Realities:** Fire service organizations operate in local political arenas. Good labor/management and customer relations are crucial to ensuring that fire departments have maximum impact on decision which affect their future.

Ongoing Issues of National Importance

7. **Leadership:** To move successfully into the future, the fire service needs leaders capable of developing and managing their organizations in dramatically changed environments.

8. **Prevention and Public Education:** The fire service must continue to expand the resources allocated to prevention and health and safety education activities.

9. **Training and Education:** Fire service managers must increase their professional standing in order to remain credible to community policy makers and the public. This professionalism should be grounded firmly in an integrated system of nationally recognized and/or certified education and training.

10. **Fire and Life Safety Systems:** The fire service must support adoption of codes and standards that mandate the use of detection, alarm, and automatic fire sprinklers, with a special focus on residential properties.

11. **Strategic Partnerships:** The fire service must reach out to others to expand the circle of support to assure reaching the goals of public fire protection and other support activities.

12. **Data:** To successfully measure service delivery and achievement of goals, the fire service must have relevant data and should support and participate in the revised National Fire Incident Reporting System. Likewise, NFIRS should provide the local fire service relevant analysis of data collected.

13. **Environmental Issues:** The fire service must comply with the same federal, state, and local ordinances that apply to general industry and which regulate response to and mitigation of incidents, plus personnel safety, and training activities relating to the environment.

Customer Service

There are opportunities, un-exercised to a great extent, for the fire service to increase its value to the community at little cost. Programs which offer citizens support in preparing for and dealing with fires, medical emergencies, and other emergency incidents are being applied by departments throughout the country, as are programs not directly connected with emergency operations.

Customer Support

The fire service has begun a process to view problems in the customer's terms, not in the fire department's terms. This involves changing the ways the fire service has traditionally used its resources by expanding the perspective of firefighters and fire service managers to include mitigation of the negative impact on humans as well as property.

Fire service agencies are in a unique position to reduce the negative consequences of emergencies by assisting in their immediate recovery from emergency incidents.

-   Immediate shelter

- Management of personal items and valuables

- Loss control techniques

- Connection to social services

- Coordinating customer support at the end of emergency operations to assist in effectively reconnecting customer lives

Customer Service

Our customers experience many urgent needs. Sometimes, they may not be able to access the correct agency for help because of the time of day or unfamiliarity with the community. Because of the relatively easy access to dispatch centers through 9-1-1, dispatchers often receive requests for assistance which do not fit the traditional mission of the fire department.

Fire agencies should develop programs which allow the quick removal of the call from the 9-1-1 lines while also providing assistance to the calling parties. This action may be accomplished at the dispatch center or fire station providdng a list of contact parties with 24-hour telephone numbers for such services as:

- Social service support

- Animal control organizations

- Mental health organizations

- Behavioral health services

Departments should develop a culture of citizen assistance which is reflected in the use of department resources to provide non-traditional support to customers in related pubic safety and community support ways. Examples of programs some departments have developed are:

- Assistance to disabled motorists (from obtaining fuel to using a cellular phone to call the "auto club")

- Hosting neighborhood blood pressure testing clinics

- Providing bicycle registration services and safety clinics

- Drowning prevention programs

Fire training and educational organizations should initiate programs to make fire departments aware of these opportunities, their value to fire departments, as well as the communities they serve, and methods for incorporating such programs into fire department operations.

Impact of Managed Care

The advent of managed care promises to reduce or control increases in the cost of health care. Accompanying this movement is the potential need for radical changes in the delivery of emergency medical services, from dispatching to first responder, to ambulance treatment and transportation.

The American fire service possesses a valuable tradition of service, unmatched resources, and a trustful relationship with the public, all of which can be used to facilitate the goals of managed care.

Fire service leaders at all levels and the elected and appointed officials to whom they report must be open to organizational and philosophical changes which will permit the successful participation of fire service personnel and resources in the delivery of emergency medical services within the framework of managed care.

As an example of taking advantage of this opportunity, one urban fire department has joined with a local hospital to be a part of the home health care system. A physician's assistant is part of the crew assigned to an ambulance. This unit is scheduled to deliver a variety of out-of-hospital services, such as immunization and sutures, to those who do not need to receive this care in a hospital setting. This saves time and money for all parties involved, while allowing the fire service to have a direct link to be a key player in the managed care arena. The future should hold many opportunities for the fire service to be a part of managed care, if we are willing to seek out and pursue responsibilities.

Marketing/Competition/Customer Education

The fire service has never had a greater need to competitively market itself and its services. The fire service must recognize the changing environment of society and develop competitive strategies for marketing its services to its stakeholders, representing a wide spectrum of key individuals, public and elected officials, and various organizations.

The fire service must move forward to remove barriers and take the needed steps to interact with the community 365 days a year. The use of non-traditional factors, such as those used by business and industry need to adopted. The fire service must be visible in the community, access the customer, promote fire safety and injury prevention messages continuously, both personally and electronically, and utilize national resources for education to a much larger degree then is done presently.

As police agencies across America are reaching out to develop neighborhood community-based policing efforts, the fie service needs to return to its roots. Historically, the local fire station has been a cornerstone of community life. The fire service must capitalize on this long-term relationship and direct this effective bond towards our marketing efforts. As a part of every fire service agency's mission statement, a statement should identify community involvement programs the department supports. One very successful marketing/community involvement program is Explorer Scouting. Many agencies have flourishing programs that should be expanded and valued by all department members.

Service Delivery

The work of the fire service continues to grow and become more complex each day. Experience indicates that budgets are being cut, staffing levels are reduced, and the ability to purchase necessary tools and equipment is hampered. It is imperative that the American fire service support the development of nationally recognized service delivery criteria that address efficiency, effectiveness, and safety.

Service delivery requires a universally applicable standard which defines the functional organization, operation, deployment, and evaluation of public fire protection and emergency medical services. These criteria must address fire department emergency service delivery, response capabilities, communications, evaluation of risks and resources, fire prevention, fire investigation, public education, and community involvement.

Our strategy should be to create nationally developed and accepted standards, applicable to all public fire and rescue organizations, no matter what their source of personnel, which will be used to provide evaluation criteria concerning the effectiveness, efficiency, safety and timeliness of response, deployment, operations, and programs.

At the time Wingspread IV was held, the National Fire Protection Association was in the process of developing a standard entitled "Fire Department Deployment" (NFPA 1200). A nationally held and valued standard could go far in reaching this emerging strategy.

Wellness

Historically the fire service has known more about the apparatus and equipment it purchases than the firefighters who use that equipment. Firefighters will continue to respond to disastrous situations that produce extreme physical and emotional consequences. Over time, these situations can affect the overall wellness of a fire service system.

Living and working in today's society will continue to be stressful and challenging. The fire service is faced with many of the same stressors as its customers. Given this continued situation, the fire service must begin to develop total wellness systems to enable firefighters to cope, develop, be safe, and survive a lifetime of responses.

The development and implementation of wellness systems should occur in cooperation with the major fire service organizations. These systems should be long-term, holistic, positive, rehabilitative, and educational. They must overcome the historic punitive mentality of physical fitness, move beyond negative timed, task-based testing and toward progressive improvement, and require labor and management to commit to a positive, individualized program with testing and private attitude results. These systems should have a holistic approach which includes fitness, rehabilitation, behavioral health, and nutrition to ensure that firefighters reach and maintain optimal wellness.

There is obvious lack of scientific research and data about the medical/emotional/occupational diseases that effect our members. This research should lead to a comprehensive wellness program that all firefighters deserve.

Political Realities Affecting the Fire Service

Some would like to think that problems and challenges facing the fire service should be above or separate from the political process. The real world tells us just the opposite is true. Decisions about almost every aspect of the fire service are affected to some degree or another by the political process.

Funding, hiring, promotions, staffing levels, scope of service delivery, emergency transportation and, in many instances, the identify of the fire chief and what he or she does are all affected by elected and/or appointed decision makers. Directly related to issues of personnel recruitment and retention pose continuing challenges for our volunteer fire service leaders.

Career fire chiefs typically have little ability to impact the political process. Often it is the employee organizations (unions) that have the best opportunity to educate elected officials concerning the fire service needs. When fire department administrators and the employee organizations have a positive, progressive labor/management process, the two groups can work together to provide a political environment in which effective and efficient service can be provided.

By sharing the same two goals of providing the best possible fire protection and emergency medical service to the citizens for their tax dollars and taking care of the physical and mental health of those providing that service, labor, and management are in the best position to influence the community and its elected officials.

A positive, progressive, official labor/management process nurtures trust between employee organizations and fire administrators. Employee organizations generally understand and participate in the political process at a much higher level than fire administrators. The volunteer fire service must learn the political process as the employee organizations have in order to survive. The labor/management process provides the glue necessary for employee organizations and fire administrators to work together for the betterment of the fire service as a whole. Labor and management, working together, ensure that the fire service will have a seat at the table when decisions are made which affect its future.

Leadership

To move successfully into the future, the fire service needs leaders capable of developing and managing their organizations in dramatically changed environments.

To compete successfully, the fire service needs smart, tough, nice, modern managers. These managers must be able to operate successfully in competitive, changing, and non-traditional environments. They must have vision, ability to predict, and effective human relations skills.

Fire service leaders will need to be recruited, selected, trained, and supported. They will need strategies to overcome the autocratic, power-focused, and controlling environments of the past. Significant value shifts in management philosophy must take place. Among the challenges fire service leaders must successfully face all firefighter safety, human resource management, effective customer service, labor/management relations, diversity of the community and the work force, and financial realities.

Fire Prevention and Public Education

Fire and emergency services must continue to expand the resources allocated to prevention and education activities that have the goal of reducing injuries and deaths from fire and other risks.

Public fire education has progressed significantly in the past twenty years. Programs now are targeted for specific audiences and specific risks, and are better evaluated. Fire safety messages have been refined. The concept of teaching people how to prevent fire and how to take proper actions in case of fire has been expanded to include teaching people how to prevent many "accidents." This concept is called "all risk prevention," "injury control," "fire and life safety," and/or "multi-hazard prevention."

Public fire and safety education has emerged as a profession. To continue the progress of the past twenty years, the profession must:

1. Develop standards for programs and messages

2. Develop more messages about the technology of detection, alarm, and automatic sprinkler systems in residential properties

3. Include education of elected and appointed officials

4. Use locally based methodologies and initiatives to educate citizens and customers

5. Build into programs a method of evaluation to determine if public education is achieving its goal of behavioral change

Education and Training

Fire and emergency services managers must increase their professional standing in order to retain credibility with the policy makers and the community at large. Such professionalism should be firmly grounded in an integrated system of nationally recognized and/or certified education and training.

The tenets of professionalism are well established; a body of knowledge; formalized education system for acquiring that knowledge; a recognition of status, service over profit; qualification of individual competency; character; and an assurance to the public of the competence of the member.

Currently, there are varying levels of education and training available to the fire service. There is a very strong system of training available through a closely linked cooperative effort of federal, state, and local training systems. There is also a loosely linked system of higher education systems offering associate and baccalaureate degrees in the fire sciences, public management, and engineering disciplines. There are fewer graduate opportunities for those seeking higher professional status. There is a wide general agreement that the federal fire programs of the U.S. Fire Administration have significantly improved since the last Wingspread Conference, and as a result, there are now more professional opportunities available.

The next logical sequence for increased professionalism lies in the following areas:

1. Skills-based knowledge simulation training, similar to that available to aviation and the military, that will establish and maintain emergency management skills. This training should be available through state and local training systems. This system is currently under development at the National Fire Academy and other training institutions.

2. A strengthening of the already well established cooperation between the National Fire Academy and state training systems, and a commensurate link through to local training systems. These conduits will increase the availability of training opportunities, while significantly reducing development and delivery costs.

3. Mid to senior level fire and emergency services managers must have college experience if recognition of their professional status is going to be maintained. Fire and emergency services managers of the future must be prepared to discuss issues, on an equal academic footing, with architects, engineers,

city managers, and health care professionals. A master's degree in a discipline of relevance to the fire service, an appropriate level of training, and line and staff experience commensurate with the responsibilities of the position, should be the minimum acceptable qualifications for a career fire chief. Recognizing that there are variances in need, opportunities, and availability, the volunteer fire services must seek to promote individuals with levels of training and education that reflect as closely as possible this aspiration.

4. Fire service managers should encourage certification through either of the traditional venues, International Fire Service Accreditation Congress or the National Board of Fire Service Professional Qualifications, or both.

Fire Protection Systems Technology

Fire related death and injuries in residential properties continue at an unacceptable level. Fire death and injury reports indicate that traumatic results affect primarily our youth and our elderly. The technology is available today that can significantly reduce those unfortunate losses.

It must be the responsibility of the American fire service to embrace the technological advancements in fire detection, alarm, and built-in automatic fire suppression systems. The fire service in general needs to be better educated about the available standards, usage, and costs of residential sprinklers and smoke detector systems. To have a significant impact, the fire service must push for increased education for the public of the benefits of a complete life safety system.

The nation's fire service should support the adoption of codes and standards at the local, state, and national levels that mandate the use of detection, alarm, and automatic fire sprinklers. New programs must be developed in partnership with the private sector to better educate the public about the realistic benefits and reasonable costs of residential fire sprinklers. The fire service must be the lead agency to ensure that codes and standards are followed.

The fire service must be educated about the technological advancements with alarm and detection systems. For example, the increased use of the new ten-year smoke detector and the carbon monoxide detector will have a significant effect on reducing fire related deaths and injuries.

The fire service must take responsibility in having programs developed and used that will promote the proper maintenance of these fire protection systems. We must learn from the past and avoid problems in the future with dead or missing batteries.

For example, one suburban fire department conducts an ongoing door-to-door smoke detector installation program. On duty fire companies strategically canvas all neighborhoods to check and/or install smoke detectors in homes. Funding for this program is provided by a local fire insurance agency. This department has incorporated 100% community residential smoke detection into its goals statement.

With increased fire service involvement and support of the technological advancements in fire detection, alarm, and suppression, a much clearer understanding of the benefits and costs of residential fire protection can be achieved.

Strategic Partnerships

The fire service must reach out to others to expand the circle of support to ensure that the goals of fire and accident prevention are reached. The fire service community can no longer operate in a vacuum and expect to serve in the changing environment (technology, politics, information).

The fire service must forge strategies, alliances, and partnerships at the federal, state, and local levels, as well as with the private sector. National partnerships that have developed, such as the "Change your clock, change your battery" program with the Energizer Battery, should be emulated at the state and local levels.

Partnerships facilitate:

1. Accomplishing common goals

2. Communication and networking

3. More effective multi-agency emergency incident response

4. Reaching political goals

Possible alliances: law enforcement, emergency management, volunteer organizations, labor organizations, community service organizations, health maintenance organizations, insurance industry, industry and business associations, professional associations with traditional affiliation of mayors, public administrators, elected officials, development of consortiums.

Data

With the proliferation and availability of data, few organizations can ignore their critical importance. For the fire and emergency service, the system for data collection and analysis is the National Fire Incident Reporting System (NFIRS). Currently, the system is under revision and is expected to include modules for emergency medical responses, hazardous materials, and wildland fire response data.

At this writing, the system does not enjoy the benefit of full fire service participation for a number of reasons. Until the fire service can produce the data equivalent and accuracy of the Uniform Crime Report (quarterly), they will continue to lag behind, never able to accurately characterize or articulate the challenges they have met or will face.

The changing roles of the fire and emergency services into an all-hazards/all-risk service delivery system, underscores the importance of the national data collection system. The development of public policy, as well as the proper management of resources, should be founded upon the critical analysis of uniform data. On a local level, the ability to access the quantity and quality of services, to measure the impact of these services, to plan for levels of needs, and to design and implement improvements are the elemental tools for the fire and emergency services manager. There is a need for an improved NFIRS system. The new system must address the following issues:

Clarify the collection requirements: rid the system of those current data elements and codes that (a) are no longer relevant; or (b) are so confusing or burdensome that they diminish the likelihood of complete and accurate data entry; or (c) can be derived from other information.

Simplify the forms: clarify and simplify the rules for completion of paper and automated forms.

Accommodate local information needs: federal, state, and local information needs are not coincident. Participation is encouraged if as many as possible of the diverse needs are accommodated in a single system.

Expand the breadth of the system of all incidents: since the introduction of NFIRS, it has become increasingly important to document the full range of fire department activities. For example, NFIRS 4.1 does not address EMS incidents. Consequently, the new NFIRS should encompass the full range of departmental activities.

Collect data relevant to incident suppression/mitigation: since the current system was not designed by the fire service or those who used incident data, parts of the data that are currently collected in the system are not used. Other parts are imperfectly designed, leading to poor utilization for analysis or prevention programs. The new system must produce summary based data and analysis, as well as task level needs analysis.

Before the millennium, every fire agency in the country should be a full participant in the state NFIRS-based reporting system. Every state should be a participant in the national system. Local governments should insist on a state system that is a participant in the national system.

While the NFIRS system will produce analyzed data locally, communities participating in NFIRS should automatically receive data, based on broad data sets, formatted in a standard format, upon which they can make strategic decisions. Special data analysis should also be available upon request of participants at minimum cost.

Data poorly produced or improperly presented is damaging. Federal, state, and local training organizations should provide training in the use, production, and analysis of NFIRS data.

There are other sources of fire data that may be useful to the fire and emergency services manager. The annual Fire Departments Survey, of Major Fire Losses, Firefighter Injury and Death statistics are all available from the National Fire Protection Association, as well as the Death and Injury Survey available from the International Association of Firefighters, are examples of this additional data. The National Fire Academy should conduct a biannual national survey to determine current staffing levels, equipment operated, fiscal data, work hours, shift patterns, training, education and inspection achievements, and other relevant fire service information.

Environmental Issues

The fire service must comply with the same federal, state, and local ordinances that apply to general industry and which regulate response, mitigation, personnel safety, and training activities relating to the environment. Environmental concerns will continue to impact the fire service and the community.

The fire service must take an active role at the state and federal levels to ensure that its interests are protected in all related environmental issues. Among the many areas that impact the fire service are laws, regulations and standards, and various response and training funding sources.

The fire service will need to develop policies that support the protection of the environment from accidental and illegal spills and releases. These policies should be planned in cooperation with federal, state, and local agencies, and the private sector, as well as environmental groups.

In support of those policies, the fire service must comply with environmental regulations as they relate to mitigation of hazardous materials, structural and wildland fires, as well as training activities that could cause unnecessary personal exposure or environmental contamination.

Conclusion

Hopefully the Wingspread Conference will inspire and challenge the fire service to address these issues in a proactive and positive manner. The fire service is an essential and integral part of the American society and must adapt its roles to meet the needs of its diverse and changing customer base. The goal and purpose of Wingspread is to stimulate discussion and creative thinking on these issues to ensure that the fire service will be stronger for its participation in the national debate on them. The participants in Wingspread IV hope that Wingspread V will document significant progress in these issues and identify new and different issues that need to be addressed.

Wingspread IV Participants

Alan V. Brunacini, Fire Chief; Phoenix, AZ Fire Department

Patrick E. Cantelme, President; United Phoenix, AZ Fire Fighters Local 493

Richard Duffy, Director of Health and Safety; International Association of Fire Fighters, Washington, DC

Harvey Eisner, Editor-in-Chief; Firehouse Magazine, Melville, NY

Dr. John Granito, Fire Protection Consultant; St. James City, FL

Raymond E. Hawkins, Director of Client Education and Training Services; Volunteer Firemen's Insurance Services (VFIS), York, PA

Mary Beth Michos, Fire Chief; Prince William County Fire and Rescue, VA

William Neville, Consultant; Neville Associates, Penn Valley, CA

Dr. Denis Onieal, Superintendent; National Fire Academy USFA/FEMA

James O. Page, Editor; JEMS Communications, Carlsbad, CA

Dennis L. Rubin, Fire Chief; Dothan, AL Fire Department

Thomas L. Siegfried, Fire Chief, Retired; Altamonte Springs, FL Fire Department

Joe M. Starnes, Fire Chief; Sandy Ridge, NC Volunteer Fire Department

Steve Storment, Assistant Chief; Phoenix, AZ Fire Department

Nancy J. Trench, Director of Fire Service Training; Oklahoma State University Stillwater, OK

Bruce Varner, Fire Chief; Carrollton, TX Fire Department

Recorders:

Mark R. Nugent, Senior Captain; Chesterfield, VA Fire Department

Robert Tutterow, Safety and Logistics Officer; Charlotte, NC Fire Department

Sally Young, Fire Department Planner; Charlotte, NC Fire Department

Facilitators:

William D. Lewis, Education Specialist; National Emergency Training Center, FEMA, MD

R. Wayne Powell, Program Chair for Fire Prevention Management; National Fire Academy, USFA/FEMA, MD

Wingspread I, 1966: Statements of National Significance

1. Unprecedented demands are being imposed in the fire service by rapid social and technological change.

2. The public is complacent toward the rising trend of life and property loss by the fire service.

3. There is a serious lack of communication between the public and the fire service.

4. Behavior patterns of the public have a direct influence on the fire problem.

5. The insurance interest has exerted a strong influence on the organization of the fire service. This dominance seems to be waning. The fire service must provide the leadership in establishing realistic criteria for determining proper levels of fire protection.

6. Professional status begins with education.

7. The scope, degree, and depth of the educational requirements for the efficient functioning of the fire service must be examined.

8. Increased mobility at the executive level of the fire service will be important to the achievement of professional status.

9. The career development of the fire executive must be systemic and deliberate.

10. Governing bodies and municipal administrators generally do not recognize the need for executive development of the fire officer.

11. Fire service labor and management, municipal officers and administrators must join together if professionalism is to become a reality.

12. The traditional concept that fire protection is strictly a responsibility of the local governments must be reexamined.

Wingspread II, 1976: Statements of National Significance

1. New criteria is needed to measure the impact of fire on the national economy and public welfare.

2. Productivity in the fire service is difficult to measure reliably.

3. The state level of government may have to make a renewed commitment in dealing with the fire problem.

4. The fire service should approach the concept of regionalization without bias.

5. There is a need for a better liaison between the fire service and those who build or design buildings.

6. A means of deliberate and systematic development of all fire service personnel through the executive level is still needed.

7. The firefighter has been suppressed by narrow education and confirming experiences on his job.

8. The problem of arson in the United States has increased to the point where it should be considered a matter of major importance.

9. Fire departments should thoroughly analyze new demands being placed upon them before accepting new responsibilities.

10. It appears that residential smoke detectors hold the most practical potential at this time for savings. The fire service should take leadership in encouraging their widespread use and proper maintenance.

11. Traditional fire services should assume more responsibility and leadership in fire loss management.

Wingspread III, 1986: Statements of National Significance

1. Society in general appears unwilling to take full advantage of the knowledge and technology which has proved effective in mitigating the fire problem.

2. Public fire safety education will not achieve its potential until it is organized in a systematic manner based on human behavior.

3. Professional development in the fire service has made significant strides, but improvement is still needed.

4. Decision makers in local government need better criteria to determine an adequate level of cost-effective fire protection.

5. The fire service should review the effectiveness of the federal fire programs of the U.S. Fire Administration and National Fire Academy to determine if they are of continued benefit in reducing the fire problem.

6. The traditional role of fire departments is changing.

7. Analyzing America's fire problem requires a more effective system of data collection.

8. The misuse of alcohol and controlled substance is a serious fire service problem.

9. There is a need for increased emphasis on firefighter safety and health.

10. Personnel management in the fire service is becoming increasingly more complex.

Appendix B
Examples of Phoenix, Arizona Fire
Department Customer Service Surveys

In April, 1996, the Phoenix Fire Department began mailing a Customer Survey to its customers who had called 911 for a fire or EMS situation. This was done by the Corporate Commmunications/ Publications Section of the department. The survey is printed in English on one side and Spanish on the other. The survey is sent with a postage paid envelope and is returned to a firm contracted by the Fire Department to compile the statistics. These statistics are forwarded to the Fire Department and will provide both positive feedback as well as constructive criticism for the Department.

The demographic information collected will also be helpful in making decisions regarding what is needed in the community now and in the future. The Department will be able to use this information to determine the effectiveness of its current programs and whether or not they are meeting the customers' needs.

The parents of the children who attended the educational intervention classes for older childred in the Youth Firesetter Prevention Program receive a self addressed and stamped evaluation card. The return rate on these has exceeded the department's expectations. The number of children who don't set another fire has been in the 97% range.

This particular questionnaire also acts as a reminder to both the parents and children as the importance of continuing to follow fire safe behaviors.

City of Phoenix

FIRE DEPARTMENT
CORPORATE COMMUNICATIONS

Our Family Helping Yours

Winner of the
Carl Bertelsmann
Prize for

November 27, 1996

Dear Citizen,

The Phoenix Fire Department recently responded to an incident which you were involved in. In an effort to gauge and enhance the level of service provided to you, our customer, we have instituted a Customer Service Survey Program.

Responses to this survey are confidential. The firefighters who responded to your incident will not be provided with any of your responses. All responses will be compiled by Shoreline Associates, an outside company who will summarize the results of the survey responses and provide recommendations for improvement.

If this letter has been addressed to a minor, a responsible adult or legal guardian should complete it. Please complete the survey as soon as possible and forward it to Shoreline Associates in the enclosed self-addressed, stamped envelope.

We thank you, in advance, for your assistance and completion of this survey. Please be assured that your opinions are valued. We hope that you were satisfied with the level of service provided by the Phoenix Fire Department. We are available for comments and/or questions at (602) 262-6002, anytime.

Once again, thank you.

Sincerely,

Alan V. Brunacini
Fire Chief
Phoenix Fire Department

Customer Service Survey
City of Phoenix Fire Department

I. Please begin by telling us a little about yourself.

1. Your age: o Under 21 o 21-30 o 31-40 o 41-50
   o 51-60   o 61-70   o Over 70

2. Your sex: o Male o Female

3. Your marital status: o Single o Married
   o Separated/Divorced o Widowed

4. Your race: o White/Caucasian o African American
   o Native American o Hispanic
   o Asian-American   o Other_____

5. The number of relatives and/or dependents currently living in the same household as you:
   o 0   o 1 to 3   o 4 to 6   o Over 6

6. Number of years you have lived in the City of Phoenix:
   o Less than 1   o 1 to 3   o 3 to 5   o 6 to 10
   o 11 to 20   o Over 20

7. The highest level of education that you have completed is:
   o Grade School   o Some High School
   o High School   o Some College o College

8. Do you have:
   a. Smoke detectors in your home?
      o Yes o No *(Go to question 8c)*
   b. How many months has it been since the batteries were changed?
      o Less than 3 months   o 3 to 6 months
      o 7 to 12 months   o Over 12 months
   c. Do you have a fire sprinkler system inside your home?
      o Yes o No

9. If this incident took place at a business, please answer the following. *(If not, go to Section II)*

   a. What type of operation does your business primarily perform? *(choose only one)*
      o Manufacturing   o Assembly   o Distribution
      o Sales Office   o Retail Store
      o Repair/Maintenance Facility o Corporate or General Office

   b. Total number of employees at this location:
      o Less than 5   o 6 to 10   o 11 to 25   o 26 to 50
      o 51 to 100   o 101 to 250   o Over 250

   c. Number of years operating in the City of Phoenix:
      o Less than 1   o 1 to 3   o 3 to 5   o 6 to 10
      o 11 to 20   o Over 20

II. Tell us a little about your experience with the Fire Department in connection with this recent incident.

1. What type of emergency was it?
   o Emergency medical assistance call
   o Motor vehicle accident
   o Smoke/fire alarm activated; but no fire found
   o Fire (inside the building) o Fire (outside of the building)
   o Car fire o Other_____

2. Please rate the level of service provided by each of the following:

	Excellent	Above Average	Average	Below Average	Not Sure
a. 911 Operator	o	o	o	o	o
b. Fire Officer(s)	o	o	o	o	o
c. Firefighters	o	o	o	o	o
d. Paramedics	o	o	o	o	o
e. Occupant Services Officer	o	o	o	o	o

3. Did someone from the Fire Department explain what was happening while the firefighters were working?
   o Yes   o No (Go to question 4)

   If so, did they explain their actions in terms that you could understand? o Yes   o No

4. If the emergency was a fire within the building; after the fire was extinguished, did someone from the Fire Department:

	Yes	No	Not Sure
a. Explain what procedures were performed and why	o	o	o
b. Help you contact family and/or friends	o	o	o
c. Provide information on how to:			
1. Inventory the damage	o	o	o
2. Value the loss	o	o	o
3. Return the building or home to a usable condition	o	o	o
4. Obtain a copy of the fire report	o	o	o
5. Salvage your remaining belongings	o	o	o
6. Contact the Red Cross or other social service agencies	o	o	o
7. Provide comments/complaint to the Phoenix Fire Department	o	o	o

5. Did any of the firefighters' actions impress or upset you?
   o Yes   o No
   If so, please briefly explain: _____
   _____

6. Were there any services that you believe should have been performed, that were not?
   o Yes   o No
   If so, please briefly explain: _____
   _____

7. Was your property left in satisfactory condition?
   o Yes   o No
   If not, please briefly explain:: _____
   _____

Please check mark your answers clearly.

Ciudad de Phoenix

DEPARTAMENTO DE BOMBEROS
COMUNICACIONES INCORPORADA

Nuestra Familia Ayudando A La Suya

Winner of the
Carl Bertelsmann
Prize for

27 de noviembre de 1996

Estimado(a) Ciudadano(a),

El Departamento de Bomberos de Phoenix recientemente respondió a una casualidad en la cual usted estaba envuelto(a). En un esfuerzo por calibrar y realzar el nivel de servicio que le proveemos a usted, nuestro cliente, nosotros hemos instituido un Programa de Estudiar el Servicio al Cliente.

Contestaciones a este estudio son confidenciales. No se les proporcionarán a los bomberos que respondieron a su casualidad con cualquiera de sus contestaciones. Todos las contestaciones serán compiladas por Shoreline Asociados, una compañía externa que resumirá los resultados de las contestaciones del estudio y proporciona recomendaciones.

Si se ha dirigido esta carta a un menor, un adulto responsable o tutor debe completar el estudio. Favor de completar el estudio lo más pronto posible y envielo a Shoreline Asocia en el sobre ya estampado y con dirección que encontrara adjunto a esta carta.

Nosotros gracias le damos las de antemano, por su ayuda y realización de este estudio. Favor de sus que se tasan tus opiniones sean valoradas. Esperamos que usted este satisfecho(a) con el nivel de servicio ofrecido por el Departamento de Bomberos de Phoenix Estamos disponible a cualquir hora para comentarios y/o preguntas al (602) 262-6002.

Una vez mas, gracias.

Sinceramente,

Alan V. Brunacini
Jefe
Departamento de Bombers de Phoenix

Estudio Para el Servicio al Cliente

Ciudad de Phoenix Departamento de Bomberos

I. Favor de empezar por decirnos un poco sobre usted.

1. Su edad: o Menor de 21   o 21-30   o 31-40
   o 41-50   o 51-60   o 61-70   o Mayores de 70

2. Su sexo: o Varón   o Hembra

3. Su estado civil: o Soltero(a)   o Casado(a)
   o Separado(a)/Divorciado(a)   o Viuda/Viudo

4. Su raza: o Blanco/Caucásico o Americano-Africano
   o Nativo-Americano   o Hispano
   o Asiático-Americano   o Otro _____

5. El número de parientes y/o dependientes que viven actualmente con usted:
   o 0   o 1 a 3   o 4 a 6   o Sobre 6

6. Número de años que has vivido en la Ciudad de Phoenix:
   o Menos de 1 año o 1 a 3   o 3 a 5
   o 6 a 10   o 11 a 20 o Sobre 20 años

7. El nivel más alto de educación que has completado es:
   o Escuela primaria o Alguna escuela secundaria
   o Escuela secundaria o Alguna Universidad
   o Universidad

8. Usted tiene:
   a. Detectores de humo en su hogar?
   o Si   o No *(Vaya a la pregunta 8c)*
   b. Hace cuánto meses se les cambiaron las baterías?
   o Menos de 3 meses o 3 a 6 meses
   o 7 a 12 meses   o Más de 12 meses
   c. Tienes un sistema de rociador de fuego dentro de tu hogar?
   o Si   o No

9. Si esta casualidad se produjo en un negocio, favor de contestar lo siguiente. *(Si no, vaya a la Sección II)*
   a. Qué tipo de funcionamiento principalmente ejecuta su negocio? *(escoge uno)*
   o Industrial   o Asamblea
   o Distribución   o Oficina de ventas
   o Tienda de menudeo
   o Reparación/ Facilidad de Mantenimiento
   o Sociedad U Oficina General
   b. Número total de empleados en esta localidad:
   o Menos de 5   o 6 a 10   o 11 a 25
   o 26 a 50   o 51 a 100 o 101 a 250
   o Más de 250
   c. Número de años de operación en la Ciudad de Phoenix:
   o Menos de 1 año o 1 a 3   o 3 a 5
   o 6 ta10   o 11 a 20 o Más de 20 años

II. Diganos un poco sobre su experiencia con el Departamento de Bomberos en relación a esta reciente casualidad.

1. Que tipo de emergencia era?
   o Llamada de asistencia medica de emergencia
   o Accidente de vehículo de motor
   o Humo/fuego de alarma activa; pero no se encontró fuego
   o Fuego (dentro del edificio) o Fuego (fuera del edificio)
   o Auto en fuegoo   o Otro_____

2. Favor de tasar el nivel de servicio ofrecido por cada uno de lo siguiente:

	Excelente	Sobre Prommedio	Promedio	Abajo Promedio	No Seguro(a)
a 911 operador	o	o	o	o	o
b Funcionarios de fuego	o	o	o	o	o
c Bomberos	o	o	o	o	o
d Paramédicos	o	o	o	o	o
e Funcionarios ocupante de Servicios	o	o	o	o	o

3. Alguien del Departamento de Bomberos le explico lo que pasaba mientras los bomberos trabajaban?
   o Si   o No *(Vaya a la pregunta 4)*

   Si así fue, le explico las acciones en términos que usted pudo entender?   o Si   o No

4. Si la emergencia fue un fuego dentro del edificio; después de que el fuego se extinguió, alguien del Departamento de Bomberos:

	Si	No	No Seguro(a)
a Explicó los procedimientos que se ejecutaron y porqué	o	o	o
b Ayudo a que se comunicara con su familia y/o amigos	o	o	o
c Proporciono información en:			
1 Inventario del daño	o	o	o
2 Valor de la pérdida	o	o	o
3 Retorno el edificio o hogar en una condición utilizable	o	o	o
4 Obtuvo una copia del reporte del fuego	o	o	o
5 Salvamento de sus pertenencias restantes	o	o	o
6 Contacto la Cruz Rojo U otra agencia de servicio sociales	o	o	o
7 Proporciono comentarios/quejas al Departamento de Bombero de Phoenix	o	o	o

5. Alguna acción de los bomberos le enojó o le impresiono?
   o Si   o No Si así fue, por favor explicar brevemente:
   _____

6. Hubieron algunos servicios que usted cree que debieron serejecutados y no lo fueron?
   o Si   o No Si así fue, por favor explicar brevemente:
   _____

7. Su propiedad quedo en una condición satisfactoria?
   o Si o No Si contesto no, por favor explicar brevemente
   _____

Phoenix Fire Department

Preschool Fire Safety Class

LETTER # 1

Month, Date, Year

Dear Mr. and Mrs. *Name*,

We hope *NAME* remembers the fire safety behaviors we presented on *DATE.* It is important that you and your child review these behaviors often for your safety.

We hope the *STICKER/MAGNET/ITEM* we are sending will remind your family to test your smoke detector regularly. Remember:

"Your smoke detector is your nose at night."

We have included information that is a reminder of the safety behaviors we talked about in the Fire Safety Class. We hope your *son/daughter* enjoys talking about this with you.

If we can be of further assistance, please call us at 262-6774.

Sincerely,

Signatures

Public Education Specialists
Phoenix Fire Department

"Smoke Detectors Save Lives - Test Your Smoke Detector Monthly"

Phoenix Fire Department
455 N. 5th. Street Phoenix, AZ. 85004 Ph. (602) 262-6774

Phoenix Fire Department

Preschool Fire Safety Class

November 27,1996

Dear (Parent Name),

We hope (Child" Name) remembers the fire safety behaviors we presented on August 3, 1996. It is important that you and your family review these behaviors often for your safety.

We are sending a little reminder for your family of the safety behaviors we learned. We hope you are still practicing your home escape plan and remember two ways out of every room.

If you have returned the evaluation sent previously, we thank you for your feed-back. If you have forgotten to complete the evaluation, will you please take a minute to do it now? It is important to our program.

If we can be of further assistance, please call us at 262-6774.

Sincerely,

Public Education Specialists   (3)
Phoenix Fire Department

"Smoke Detectors Save Lives - Test Your Smoke Detector Monthly"

Phoenix Fire Department
455 N. 5th. Street Phoenix, AZ. 85004 Ph. (602) 262-6774

Phoenix Fire Department

Preschool Fire Safety Class

Dear Parent,

We hope your child remembers the information we presented at the Fire Department Fire Safety Class.

We are sending you this questionnaire to help us evaluate our preschool program.

1. Has your child played with matches or lighters since the class?

    _____YES   _____NO

2. Which of the following behaviors can they show or tell you about?

    Crawl low under smoke?   _____YES   _____NO

    Stop, drop and roll?   _____YES   _____NO

    Two ways out?   _____YES   _____NO

    Safe meeting place?   _____YES   _____NO

    Test your smoke detector?   _____YES   _____NO

    Cool a burn?   _____YES   _____NO

Please return this to us in the stamped, self-addressed envelope provided.

Thank you for your help in our program evaluation.

Public Education Specialists   (2)

"Smoke Detectors Save Lives - Test Your Smoke Detector Monthly"

Phoenix Fire Department
455 N. 5th. Street Phoenix, AZ. 85004 Ph. (602) 262-6774

APPENDIX C
A Step Back Helps Us Move Forward
Doug Hamp

On December 5, 1995, the city council of the City of El Paso de Robles, California agreed with the fire chief and firefighters of the Paso Robles Fire Department to eliminate the department.

Here's what happened. The Paso Robles Fire Department, like many other departments, had a need for additional staffing, new equipment and facilities, and an increase in its general operating budget. Sound familiar?

Taking an analytical approach to accomplishing those objectives, the department created an objective-based plan in which target issues were identified dealing with specific programs, capital outlay needs and the like. Each need or program was presented with anticipated cost, and when possible, options for meeting the need.

The plan took about one and a half years to complete. Everyone from the chief to the newest firefighter asked questions and provided ideas. As an organization we also asked the city council to help us in identifying who we were, what we will look like in the future and what level of service is reasonable.

During our self-analysis, we learned many interesting things about ourselves and the fire service in general. As the chief, I was continuously bothered that our plan was incomplete. There was something wrong. It was one of those projects that makes you continuously ask "What if?" or "How about?" In other words, it was a proposal without an end.

In January 1995, the Objective Based Plan was submitted to the city council for consideration. After reviewing the plan, and based on the urgency of local needs, the council authorized the immediate hiring of four additional firefighters. However, the council asked for further clarification of certain elements of the plan. A council member was identified as a liaison to work with the chief and the firefighter's association and report back to the council by December.

During this period, there was still a nagging feeling that something was missing in the plan. The entire membership and I stepped back and re-considered the plan as if we were not fire personnel. This alone was an amazing accomplishment. We agreed to look at the plan as if we were the city manager, a city council member or a local citizen.

One way we measured the community's perception of our department's work was that periodically, when a citizen, council member or total stranger came into the Fire Administrative Office, I asked the visitor to take a couple of minutes to draw on an easel his or her view of the department. The results showed our organization and the entire fire service as one-dimensional fire and EMS providers, with no product. The drawings also showed "fire" as the nucleus of our existence.

Generally, in most departments, the total fire call load in a given year is 4-7%. How many make the paper or other local media? Even if they do, what's the product? How expensive is your service? From the point of view of a city manager, council member or taxpayer, is this cost justified? Remember, you're not looking at this as a firefighter. And yes, at any moment we know that a major fire can occur.

If you're the manager of the water, streets, police or recreation department, can you understand why the fire department should get more when you have the same needs? The fire department should also ask: Does the community expect good roads? Water that tastes good? Places for children to play sports? Safe streets? And so on.

Lets step back once more and look at who and what we really are. If you can't see it easily, think about these incidents: The Oklahoma City bombing; floods in the South, North, East and West; earthquakes and tornadoes; hurricanes; hazmat incidents; train, airplane and automobile accidents with multiple or mass casualties; wildland fires; and people who must be cut from entrapment areas.

Who's the first responder? We are! The fire service, simply because of sheer numbers if for no other reason, is the most readily available, best trained and best oriented toward emergency response. Most importantly, notice that fire and

EMS are only two of our missions, not the center of our existence. Yet we continuously say they are, and that's the problem.

If we don't educate the community about safety, and change the way we do business, don't expect to see any improvement now or in the future.

Emergency service is what we provide! Why not say it? The newly formed Paso Robles Department of Emergency Services is a true emergency organization. First we identified the major potential threats to our community: earthquake; hazmat incidents; mass casualty by rail, air or highway; flood; and yes, fire and emergency medical incidents.

The revised Objective Based Plan provided a new focus for the department. As an organization, we identified a very valuable product as the new nucleus: community preparedness. In response, we created a Reorganization and Production Matrix. The matrix identifies a budget year and a specific potential threat that the department will concentrate on. For example, this month the department is hosting its first Disaster Awareness Fair, focusing on earthquake preparedness. Members also will visit local schools to educate children on how to prepare for an earthquake and how to behave during and after one.

Simultaneously, we will be contacting the business community and local industry on developing emergency plans for earthquake preparedness and recovery. We will contact our citizens at home through neighborhood readiness programs and community-based exercises, testing our preparedness to respond and recover as a community.

Each year, the focus will be on one of the other identified community risks. Each Disaster Awareness Fair will involve demonstrations, information booths and equipment displays from emergency response agencies, public utility companies, industry, professional and business representatives, and the offices of FEMA and state and county emergency services.

One future possibility is teaching our school-aged children about the dangers of household hazmat; for example, don't mix household cleaners to make a stronger cleaner, and don't use gasoline to clean auto parts. If adults had had this opportunity as youngsters, we might have prevented many accidents.

Again, industry and business owners will receive information and invitations to send to their employees to a short training session hosted by the department on the storage and safe handling of hazmat. A major factor is that the city council and taxpayers will know what effort is being applied, how tax dollars are being used and what the expected outcome should be.

Currently, each firefighter is receiving certified instructor training. Every program presented will be professionally offered, and quality is the bottom line. Already, we have received requests from various elements of the community to join the department's "Community Volunteer Specialist" program, which would use local talent for specific programs. Our goal is to make the entire community an extension of the department.

The new Paso Robles Department of Emergency Services will have a product, be accountable and develop community-based support. A member of our organization will not just have a certain number of years in the fire service, but will be an emergency response specialist in the community's potential threats, an educator and an employee with marketable skills and strong prospects for career advancement.

Traditional fire department training programs continue, but there are now specialized training programs for our threat potentials. We still conduct code enforcement inspections and the like, but every program we participate in is focused on our new mission.

We still drive fire engines, work out of fire stations and dress like firefighters, but we plan on looking different in the future. We stepped back and are moving forward.

Reprinted with permission from *Fire Chief*, Vol. 40, No. 4, April 1996.

9781482606492